Lecture Notes in Physics

Edited by H. Araki, Kyoto, J. Ehlers, München, K. Hepp, Zürich
R. Kippenhahn, München, D. Ruelle, Bures-sur-Yvette
H. A. Weidenmüller, Heidelberg, J. Wess, Karlsruhe and J. Zittartz, Köln
Managing Editor: W. Beiglböck

320

Donald Coles (Ed.)

Perspectives in Fluid Mechanics

Proceedings of a Symposium
Held on the Occasion of
the 70th Birthday of Hans Wolfgang Liepmann
Pasadena, California, 10–12 January, 1985

Springer-Verlag
Berlin Heidelberg GmbH

Editor

Donald Coles
California Institute of Technology, 1201 E. California Blvd.
Pasadena, California 91125, USA

ISBN 978-3-662-13679-9   ISBN 978-3-540-46061-9 (eBook)
DOI 10.1007/978-3-540-46061-9

2158/3140-543210 – Printed on acid-free paper

Hans Wolfgang Liepmann, c. 1980.

A view of Hans Wolfgang Liepmann through the GALCIT 17-inch shock tube, c. 1961.

Editor's Preface

On 10-12 January, 1985, a symposium called "Perspectives in Fluid Mechanics" was held at the California Institute of Technology in Pasadena, California. The occasion was the 70th birthday of Hans Wolfgang Leopold Edmund Eugen Victor Liepmann. More than 350 persons attended the symposium. Sixteen invited papers were presented, including three papers at a popular technical level as well as the dinner address by Susan Kieffer. Financial support was provided by TRW, Inc., the Hughes Aircraft Company, the NASA Ames and Lewis Research Centers, the Office of Naval Research, the National Science Foundation, and Caltech. The symposium was organized by J. Broadwell, D. Coles, P. Dimotakis, A. Roshko, and B. Sturtevant, assisted by a distinguished advisory committee. Arrangements were coordinated by the Caltech Development Office.

Hans Wolfgang Liepmann's professional career has centered on his position since 1939 as faculty member at the California Institute of Technology and, from 1972 to 1985, as director of GALCIT (Graduate Aeronautical Laboratories). We could list Liepmann's honors and awards, culminating in the U.S. National Medal of Science, but we prefer to let the present volume speak for itself. Liepmann's choice of research fields has always been wide-ranging and has often anticipated the development of new technologies. He and his students were already publishing papers on boundary-layer stability and transition in 1940, on turbulent shear flow in 1943, on transonic flow and shock waves in 1944, on surface friction in supersonic flow in 1946, on aircraft buffeting and other stochastic problems in 1947, on rarefied gas flow in 1956, on magnetohydrodynamics and plasma physics in 1957, on the fluid mechanics of liquid helium in 1968, on the chemistry of turbulent mixing in 1976, on active boundary-layer control in 1979.

Liepmann is a superb teacher. He is noted for delegating responsibility -- and credit -- to able students, so that their own careers have the strongest possible beginning. Ten of his first fifteen students are members of the U.S. National Academy of Engineering, and two are also members of the U.S. National Academy of Science. Many of his more than 60 Ph.D. students are senior faculty members at leading universities or have positions of major responsibility in industry and in government laboratories. Many hundreds of undergraduate and graduate students at Caltech have taken Liepmann's courses in thermodynamics, gas dynamics, stochastic processes, and other subjects and have propagated Liepmann's style, especially his unvarying pursuit of clarity and excellence, to far places.

The papers in this festschrift reflect Liepmann's wide interests in science. Although a few of the manuscripts were ready at the time of the symposium, several others had to be produced by transcription from a tape recording, followed by extensive revision by the author and editor. A few authors were not able to make time in their busy schedules to complete their contributions. It was originally intended that the papers would be published in a special issue of an archival journal, and some time was required to establish that this plan was not practical. Springer-Verlag has generously agreed to make the proceedings available in their series "Lecture Notes in Physics" as a significant addition to the scientific literature. The publisher, editor, and referees share the view that the contributions published in this volume are not and will not soon be out of date.

A symposium dedicated to the career of a leader in a field can be an effective vehicle for exchange of information and ideas. There is a general atmosphere of comradeship, community, challenge, and compatibility with the ambience of the scientist being honored. The lectures can provide a valuable demonstration of the way that various senior research figures function in the uncertain area where strategy merges with tactics and knowledge merges with conjecture. The participants in the symposium hope that this published record will preserve this atmosphere, including especially exposure to unfamiliar problems that can stretch the interest and imagination of the audience.

Donald Coles
29 July 1988

Contents

Methods for Exploring the Large-Scale Ocean Turbulence

Walter Munk
Professor of Geophysics
University of California, San Diego, California 92093

I am pleased that the organizing committee has seen fit to ask a sailor to join this celebration in honor of the Theodore von Karman Professor of Aeronautics. I am inspired to recount my version of how von Karman invented the von Karman vortex street, perhaps the best-known construction that bears his name. It was in Göttingen in 1911, and Prandtl had assigned to a candidate named Hiemenz the task of measuring the pressure distribution around a cylinder immersed in a steady flow. When Hiemenz attempted to make the pressure measurements, he found to his annoyance that they fluctuated in time. He went back to Prandtl and asked, "What shall I do?," and Prandtl said, "Maybe the cylinder isn't smooth enough; you better polish it." And so Hiemenz polished it, to a German degree of perfection, but the situation persisted. As von Karman told the story, he would walk by the laboratory every morning and ask Herr Hiemenz, "Does it still oscillate?" ("Wackelt es noch?"), and Herr Hiemenz would say, "Ja, ja, Herr Professor, es wackelt noch." So Herr Hiemenz would polish it again, and the daily ritual was repeated, with Karman saying "Herr Hiemenz, wackelt es noch?" "Jawohl, Herr Professor, es wackelt noch!" After a while, Prandtl suggested that perhaps the boundaries of the tank were not sufficiently smooth; this took further time, and the situation was unchanged. One day Karman got tired of the daily ritual. On a Friday he went home and said, "I really have to think about this," and he came back on Monday with essentially a completed paper on the subject of the vortex street. I would be more comfortable in telling you this tale if it did not have an oceanographic analog, shown in Fig. 1.

The upper portion of Fig. 1 shows the ocean circulation in the north Atlantic[1] as we were taught when I first came in contact with oceanography. There is a series of streamlines going smoothly around a big gyre in the sub-tropical Atlantic. Where the streamlines are crowded the velocities are high. It was vaguely understood that this circulation was the result of a wind torque between the easterly trades and the westerly winds. The east-west asymmetry is well understood to be a result of the rotation of the earth. There were some difficulties with this simple picture of a steady circulation. Occasionally people reoccupied a station that had previously been occupied (thus violating the first law of oceanography: never take a measurement over again). They would find that the results were different from what they had been before. However, we oceanographers always have a sufficient number of defects in our instruments to be able

DEPTH (M) of 15° SURFACE

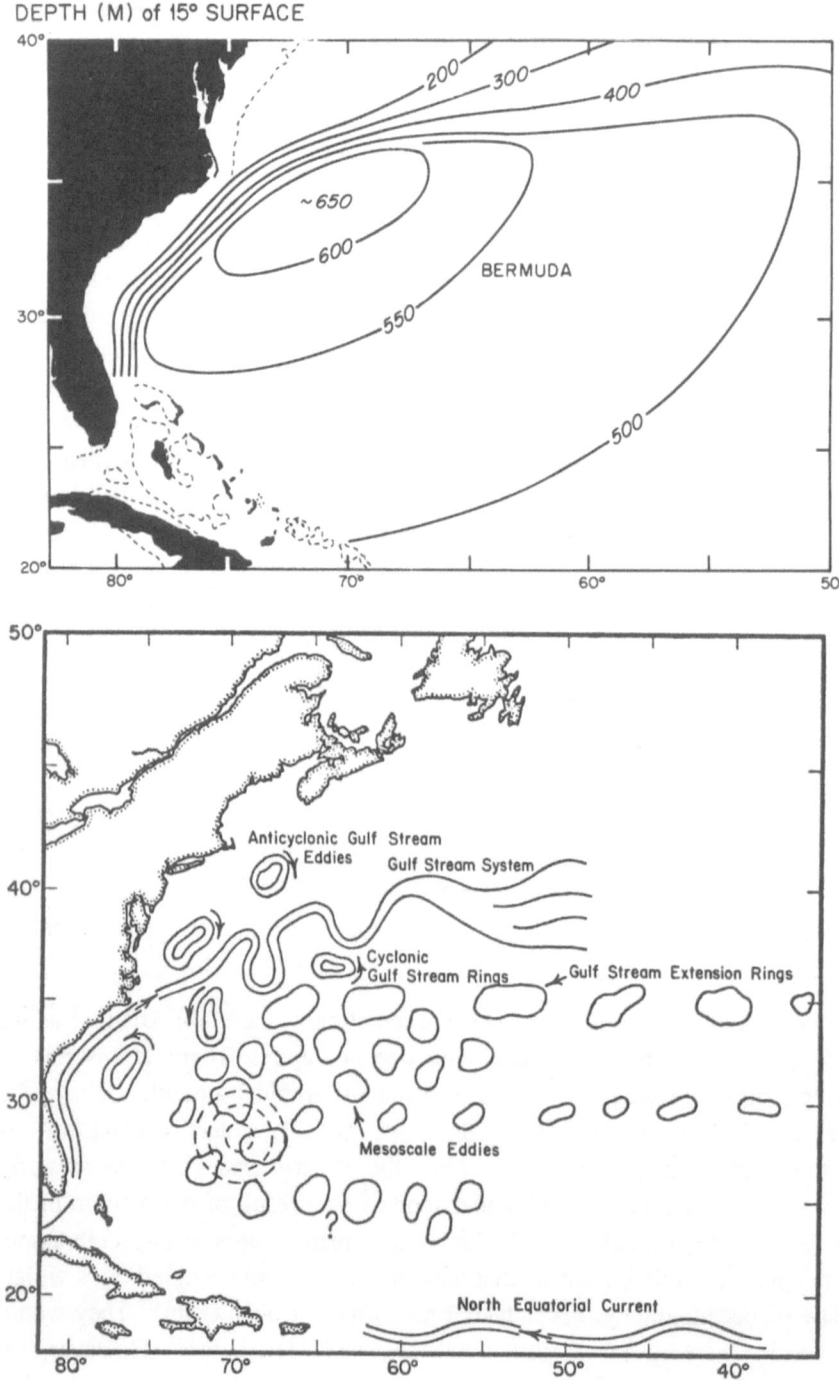

FIG. 1. *Top.* The mean circulation of the Atlantic[1] as indicated by streamlines. Crowding of streamlines indicates a high velocity. This simple pattern corresponds to our general concept of ocean circulation as the concept existed 40 years ago. *Bottom.* A cartoon of the North Atlantic circulation at any given time. The Gulf Stream meanders in space and time and the ocean is filled with mesoscale eddies.

to blame some malfunction; the ocean wiggled because we had failed to polish our instruments.

The situation became unbearable when an English oceanographer, John Swallow, invented a simple, elegant instrument known nowadays as the Swallow float. This is an aluminum tube weighted so as to be slightly heavier than water at the surface. Since aluminum is less compressible than seawater, the float eventually reaches some depth and stays there in neutral equilibrium. Swallow wanted to confirm the classical view of ocean movement, and he placed an instrument in a position north of Bermuda, where (as everyone knew) it would move to the southwest at one centimeter per second. Swallow had an acoustic means for probing for the location of the instrument. There was a transponder on the instrument, and one could follow the float from shipboard for a month or more. That was the plan. Well, the float, instead of going to the southwest at one centimeter per second, went to the east at ten centimeters per second. Even with large limits of experimental error, this result was unacceptable as a confirmation of the theory. What was even worse was that Swallow did not pay enough attention to the first law of oceanography. He placed two such instruments about 30 kilometers apart. They were, of course, supposed to float together, and it would be easy for one ship to keep track of both. Well, whereas the first floated east at ten centimeters per second, the second floated north at an equivalent speed. The whole picture of a uniform smooth circulation collapsed.

Today we regard the ocean very much more like the sketch[1] in the lower portion of Fig. 1. The ocean is filled with eddies; the Gulf Stream meanders; there are big changes in time and in space; and, what is more important, the fluctuating components associated with these eddies contain 99 percent of the kinetic energy. The top figure might still constitute a reasonable 5-year average, but it has nothing to do with what goes on at any given moment. It is almost incredible that oceanographers should have held on to the view in the top figure for so long. But so sure were we (before Swallow's experiment) of this kind of picture that we put out pocket handkerchiefs during World War II that showed downed pilots exactly how they would drift, and what to do about finding a haven. We did not tell the pilots that this picture is an average over a 5-year period, as it was not our intention that the fliers were to make a Lagrangian-particle experiment over quite so long a time. I have such a handkerchief here, and I will lend it to Hans. I have to add that this is a short-term loan. I gave this handkerchief to my wife when I asked her to marry me.

Now the scaling of eddy formation in the ocean is a different matter from the turbulent scaling that we heard discussed this morning by Professor Narasimha. The two basic facts about the ocean are that it rotates and that it is stratified. The rotation is generally measured in terms of the Coriolis parameter $f = 2\omega\sin\theta$, which is twice the rotation rate of the earth times the sine of latitude. At Caltech, near latitude 34 degrees, that is a frequency of about one cycle per day. The stratification is generally measured in terms of the frequency that a floppy balloon would have if it were filled with water at some given depth and then vertically displaced. It would oscillate with a frequency $N = [- (g/\rho)\, d\rho/dz]^{1/2}$ radians per second. In a typical ocean environment that is one to five cycles per hour, or 20 to 100 cycles per day. In the sense that the stratification frequency is much larger than the rotation frequency, you might say that the ocean is more stratified than it is rotating. This comparison has an interesting implication, because early theoreticians of the great Bergen school, V. Bjerknes and his son Jack, worked for many years on a fluid that was rotating but (for the sake of simplicity) was unstratified. In some sense, that simplification put the main emphasis backwards. We have two parameters; f, one cycle per day, and N, 20 cycles per day; two basic frequencies. The length scale that goes with rotation is the radius of the earth, a = 6370 kilometers, and the length scale b that goes with stratification is commonly defined by $N = e^{-z/b}$, since the ocean is most stratified near the surface and least stratified near the bottom. This gives a length scale b = 1 kilometer. The diameter of the eddies in Fig. 1 scales like $2\pi b\, N/f$, and is 100 kilometers in the oceans. In the atmosphere, where the stratification scale is more like 10 kilometers instead of 1 kilometer, the typical eddy size is 1000 kilometers, which is recognizable as the typical scale of a storm. The time scale of these eddies goes like a/Nb, which is about three months in the ocean and four days in the atmosphere. You see that this argument properly scales the main eddy dimensions in the ocean and in the atmosphere and shows why the two should be so different.

The discovery of eddy structure in the ocean came as a great shock, in more ways than one. How were we to sample the ocean adequately for quantities that changed vitally once every few months and that had length scales of the order of 100 kilometers? For an ocean acre (1000 kilometers times 1000 kilometers) it would take 200 days to sample adequately for 100-kilometer eddies. With a normal oceanographic ship, which costs 10,000 dollars a day, that is a very expensive operation. What is worse, in 200 days the situation has changed. So there is a real problem in attempting to sample mesoscale eddies with traditional means. At about that time, Carl Wunsch from MIT and I proposed[2] that remote sensing with acoustics might be one way to achieve adequate space-time resolution. The key is that the ocean is an excellent propagator of sound. One stick of dynamite can be heard at a distance of 1000 kilometers.

Figure 2 depicts an experiment that was performed southwest of Bermuda a few years ago[3] by a group of people whose names appear in the reference. At the bottom left is a typical ocean sound channel, a plot of sound velocity against depth. The outstanding feature is the minimum at a depth of about one kilometer. The minimum is well understood. The sound speed varies with temperature, salinity, and pressure. The sound speed increases upward near the axis because the ocean gets warmer. It increases downward away from the axis because the pressure increases with depth. The result is a minimum, which forms a wave guide. In the language of ray optics, rays that move upwards are bent downwards, and vice versa. At the bottom right in the figure is a

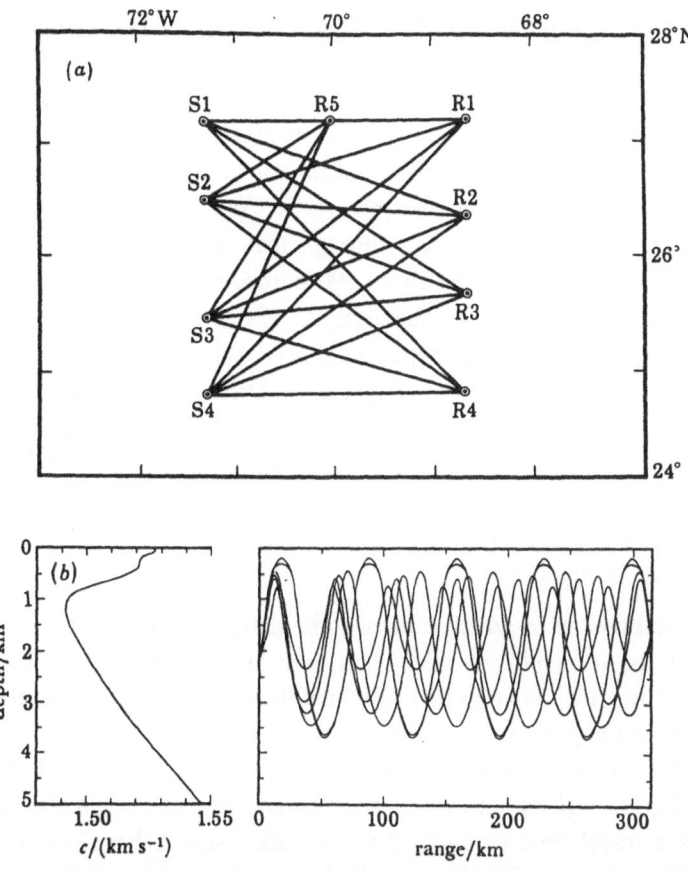

FIG. 2. A tomography experiment[3] conducted in 1981. The upper panel shows the locations of four sources and five receivers, with all possible source/receiver paths. The left bottom panel is a plot of sound speed against depth. The right bottom panel shows some representative ray paths between source S1 and receiver R3.

reasonably realistic picture[4] of ray propagation between a source and receiver separated by 300 kilometers at a depth of about 1.5 kilometers. Different arrival times are associated with different rays. The earliest arrivals, perhaps surprisingly, are the steepest ones; although they have farther to go they spend most of their time in high-velocity zones. The latest arrival is the axial ray.

Our proposal was the following. Suppose we have a warm eddy in the upper part of Fig. 2. Because the speed of sound is larger in a warm eddy, the ray traveling through the warm eddy should arrive a little earlier than it would in the absence of the eddy. When we put numbers in for a typical situation, a single eddy might give an advance by about 0.2 second, very easily measured. For a shallow eddy, only the steep ray path would go through it and come in early, whereas the flat ray path would go beneath it, and would not come in early. Clearly, the pattern of perturbation in travel time can give an indication of what is going on. We had four sources and five receivers, and we measured the perturbation in travel time from each source to each receiver. Thus we had a total of 4 times 5 times about 10 multipath arrivals, or 200 distinguishable arrivals with which we could work to find out what is going on in that ocean volume. Think of it as a 5-kilometer-deep slab, 300 x 300 kilometers in area. We measure the average sound speed along 200 rather complicated curves through that slab. The problem that we called ocean acoustic tomography is: given those 200 numbers, how can the data be inverted to produce maps of sound speed as a function of x, y, z? Since sound speed in this context means mostly temperature, the experiment produces maps of temperature as a function of x, y, z. The main advantage over the traditional method is that the information goes up geometrically, like the product of sources and receivers, whereas in traditional moorings the information goes up linearly, like the total number of moorings. This is a considerable advantage. However, as we have since learned, not only does the information increase geometrically with the number of moorings but it also decreases geometrically with the failure of moorings.

There were several questions as to whether or not this method would work. First, can individual ray arrivals be resolved? If and when they are resolved, can they be identified, so that we know how they have in fact weighted the ocean column? We need that information for the inversion process. And finally, do they remain stable over long periods of time, so that we can really work with time series? The answer to all three questions has been yes. Figure 3 (left) shows an observed arrival pattern[5]. A distance of 300 kilometers means a travel time of about 200 seconds. Notice that the earliest arrivals come at about 208 seconds, and the latest at 210 seconds, so that the dispersion over this distance is only 2 seconds. Nevertheless, there is a series of peaks in the

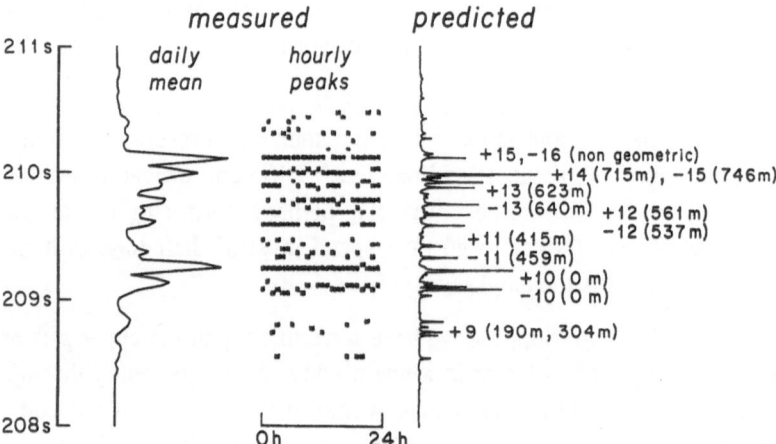

FIG. 3. A comparison of daily and predicted travel times[5]. Rays are identified by the number of turning points, by whether the launch angle is upward (+) or downward (-), and by the depths of the upper turning paths.

arrival pattern. The plot next to it shows the peaks for each hour of the day. The peaks do remain stable and identifiable. On the right side of Fig. 3 is a WKB-computed picture of when we would expect these arrivals according to ray optics. We can in fact identify the arrivals and know which ones come in at what time. The method is much like seismology applied to the ocean.

Are the results stable? Figure 4 shows the arrival pattern, where time is now plotted from left to right. Successive plots are for successive days for a total of 100 days. The pattern is in fact stable and recognizable.

The identification problem that I mentioned previously had to do solely with arrival times; steep rays come in early, flat rays late. We have recently used some simple vertical arrays so that we could also measure the angle of incoming rays to get an independent check on identification. Figure 5 shows a picture of one particular ray that

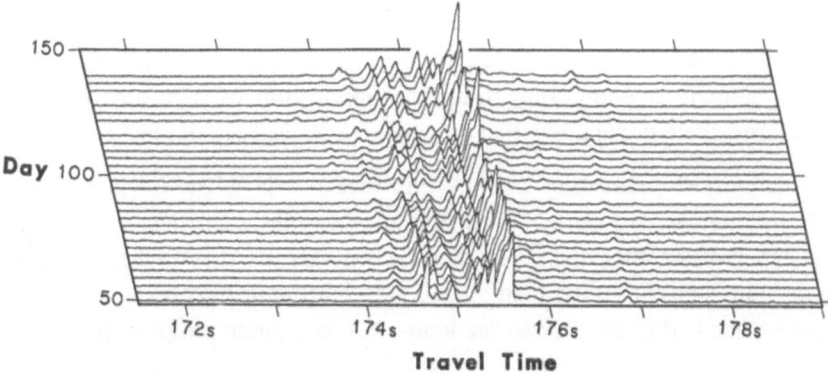

FIG. 4 The mean daily arrival pattern for a period of 100 days.

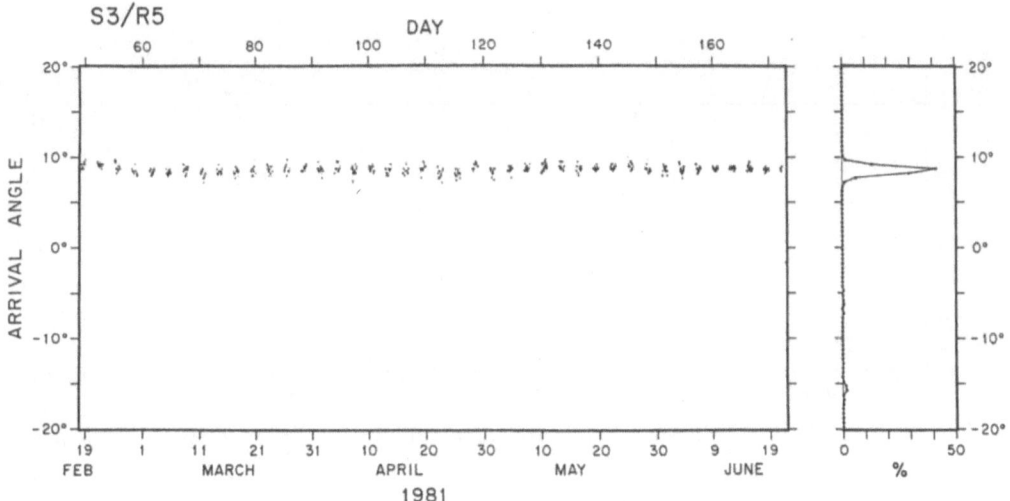

FIG. 5. The inclination at the receiver for a selected ray.

happened to arrive with an angle of about 8 1/2 degrees. The result of that investigation has been that the computed and measured arrival angles have always been consistent.

Recall that in my first figure I showed a meandering Gulf Stream. Instead of having the ideal steady streamlines of the past, it wanders. It wanders not only in space but also in time. We might therefore expect that a transmission across the Gulf Stream should have travel times that wiggle in time, because the Gulf Stream separates cold water to the left from warm water to the right. If a meander is displaced northward, as shown in Fig. 6, there is more warm water along the path, and the arrival should be sooner. This is a picture of measured changes in travel time over a period of two months. There is a change of a total of about 0.8 seconds from the shortest travel time to the longest travel time. We were fortunate to be able to compare this observed measurement of acoustic travel time with the position of the Gulf Stream as obtained from satellites. The dots give, on a similar scale, the position of the edge of the Gulf Stream relative to the line along which the acoustic transmission took place. In this instance the transmission was over a 2000-kilometer path. The total acoustic power that is transmitted in our work is about 10 watts, so we can do very well over very large distances. The agreement is good, suggesting the application of acoustic means to measure the very-large-scale fluctuations that are characteristic of the ocean.

The intellectually most interesting part of this research is what the geophysical community calls the inverse problem. Given 200 arrival times each day, can they be converted into a series of weather maps? In medical tomography (from where we stole the word), if x-rays are sent through a man's skull along different directions by rotating source and receiver, each direction gives an image. A computer program puts these

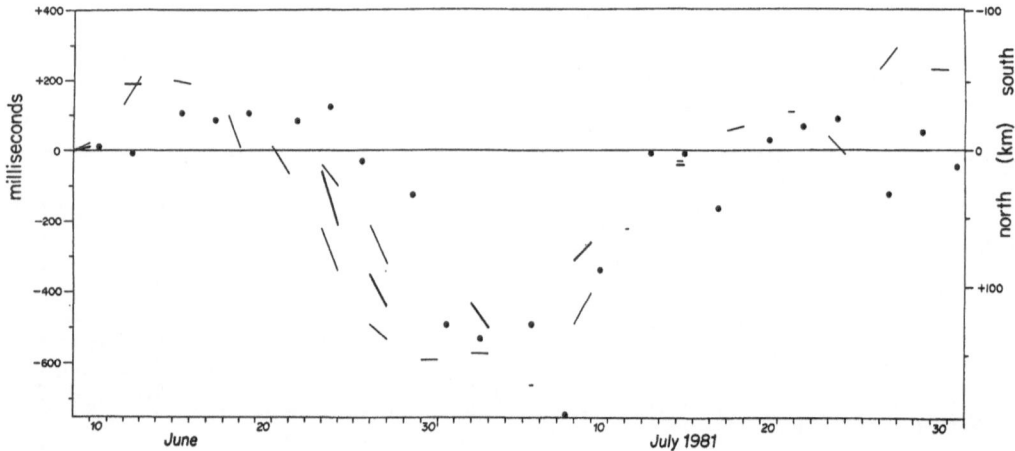

FIG. 6. Travel time between source and receiver at a 2000-km distance over a path crossing the Gulf Stream[5]. Line segments designate departures in travel time from an arbitrary mean. Dots give locations of the edge of the Gulf Stream as measured from satellites.

together into a single optimum picture. We have the same question: how do we put these 200 different paths together into a single picture? Of course, the basic statement of the problem is very simple. The total travel time is the reciprocal sound velocity, or the sound slowness, S, integrated along the path. To make the problem as simple as possible, take the ocean and break it up into a series of j blocks. Represent the sound slowness in each block j by S_j. Then the total travel-time delay Δt_i for ray i will be the sum of the perturbations of sound slowness in block j multiplied by the distance R_{ij} that ray i has traveled in block j; $\Delta t_i = \sum R_{ij} S_j$. The inverse problem is solving for the perturbation in sound slowness. We need a formula whereby the sound-slowness perturbation in block j is a linear sum of the observables; $\Delta S_j = \sum R_{ij}^{-1} \Delta t_i$. The perturbations in travel time of ray i are each multipled by an optimum weight, which in some formal sense is the inverse of the R_{ij} matrix.

Generally, the way that these problems are formulated involves more unknowns than observables, more boxes than rays. The situation is ambiguous. We are familiar with the case of 200 equations and 200 unknowns, a problem that is just determinate. We are equally familiar with the overdetermined problem, where we use least squares to solve for fewer unknowns than observables. Many people are less familiar with the remaining case, which is the underdetermined problem. It is formally very similar to the least-square problem, but it is ambiguous, and some hypothesis is needed in order to remove the ambiguity. We are using the simplest imaginable hypothesis; we are asking for the least wiggly ocean weather map that is consistent with the 200 observations, given the uncertainty in each observation. In effect, we are assuming that the ocean has a red spectrum, with more energy at low wave numbers. Given that hypothesis, the solution is no longer ambiguous. In terms of a theological problem, we are following the theology I learned in China; the sea is red. We can accept that as a statement of the spectrum of disturbances in the ocean and on that basis carry out our inversions. It is interesting that the medical profession adopts a different hypothesis. They consider a man's skull to consist of uniform fabrics separated by sharp boundaries. If we adopted that theology of the ocean, as is done by enthusiasts for frontal systems, we would get quite different maps. We really should go through the exercise of using the same data to produce maps under different theologies.

Figure 7 shows some early results (which are not as good as they should be) from the 4-source, 5-receiver experiment southwest of Bermuda. These are maps of sound speed at a depth of 700 meters at 3-day intervals, so they are snapshots of the ocean. The results are poor during days when the moorings to which our instruments were attached were leaning over because of strong currents. Furthermore, we found that the precision of our resolution for separating nearby arrivals was not really sufficient, because our acoustic sources did not have enough bandwidth. We have since improved the situation, and today we can get better results. In any event, we were able to produce these maps of eddy-like features. The maps are at 3-day intervals, which is unthinkable if traditional methods are used. We can think of the acoustic method as using a probe

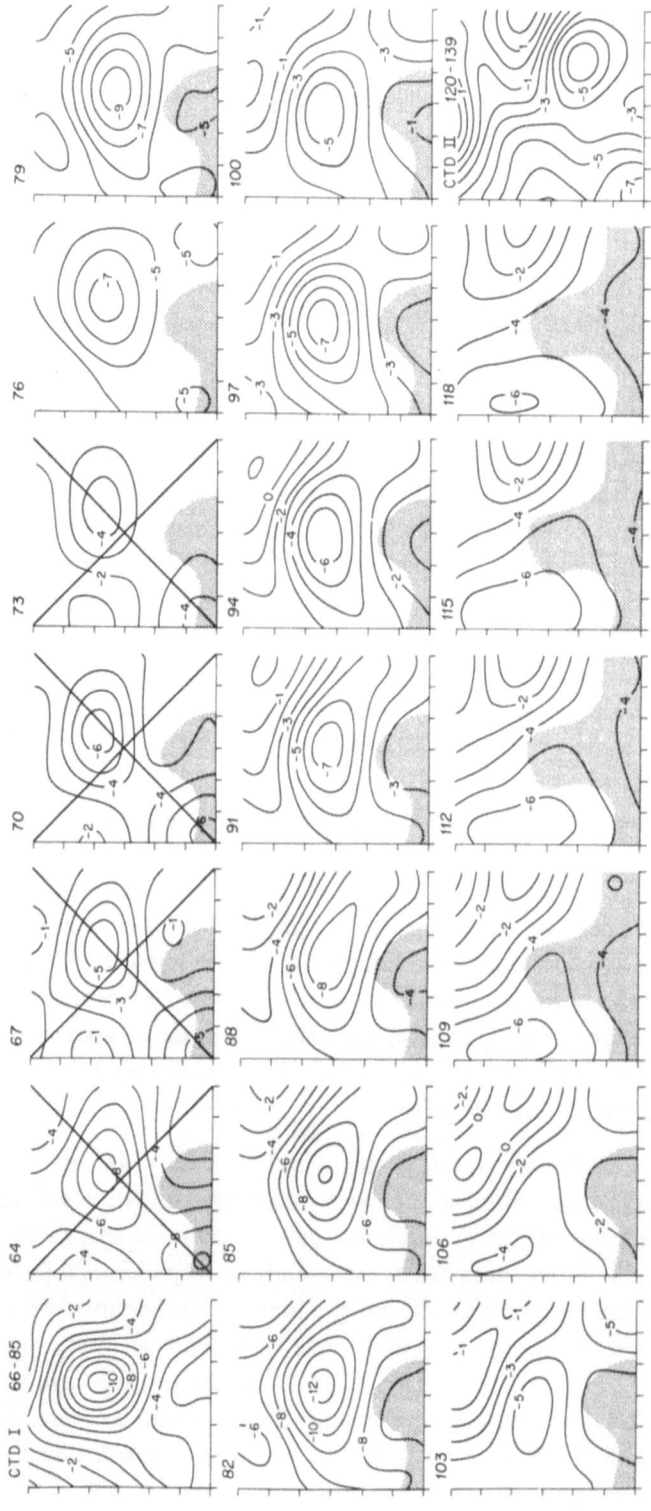

FIG. 7. Contours of sound speed at 700-m depth as calculated from the acoustic measurements of Bruce Cornuelle[3].

that moves at 3000 knots. One of the few things that has not changed since the earliest days of oceanography is the speed of oceanographic vessels. It was 12 knots during the days of the Challenger expedition, 100 years ago, and it is about 11 knots today.

Carl Wunsch and I, who were partners in this enterprise, think that the best use of such an acoustic method is in concert with satellite observations. Figure 8 is a record of the path of the the satellite SEASAT over the north Atlantic over a period of 10 days. The grid size is quite satisfactory for resolution of the eddy structure. Probably the best quantity to measure from a satellite, if you are interested in the large-scale eddy structure in the ocean, is the elevation of the sea. By measuring the time required for an electromagnetic pulse to travel to the ocean surface and back, altimeters have now achieved a precision of a few centimeters. The typical eddy signature is 10 centimeters. The principal advantage of satellite altimetry is excellent horizontal resolution. But there is no depth capability whatsoever, because the electromagnetic waves associated with such observations penetrate only a few centimeters into the water. We think that a combination of the satellite capability with an underwater acoustic scheme would constitute a good strategy for making such measurements.

We have since used the acoustic technique reciprocally, by using co-located sources and receivers to measure travel time from A to B and from B to A. The difference of

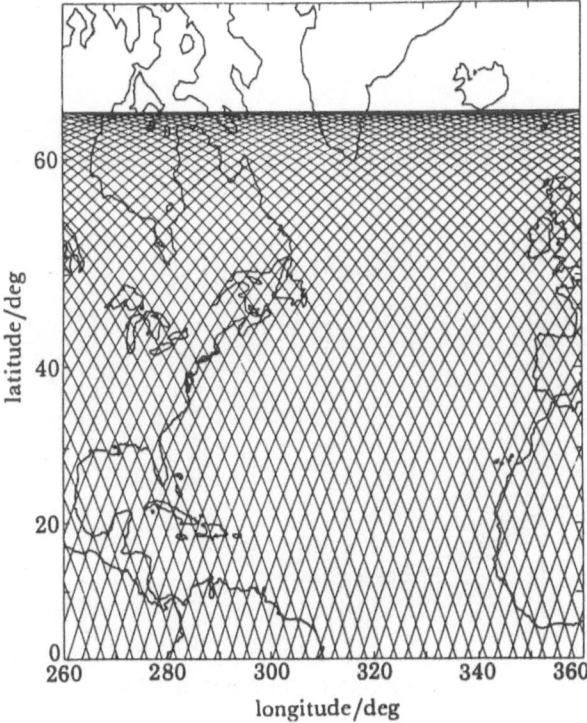

FIG. 8. Satellite paths over the Atlantic over a period of ten days[2].

the two is a measure of velocity. We can go through the same formalisms and produce maps of currents. Our plans for 1986 are to set up an array with a scale dimension of 1000 kilometers just east and north of the Hawaiian islands (Fig. 9) and to attempt to measure the variability of circulation in the ocean on a climatological scale. The array will not resolve the eddies. Moreover, the problem is nonlinear, in the sense that as the ocean changes, so do the ray paths. The simplest linear analysis is to interpret changes in travel time in terms of changes in sound speed along the undisturbed path. The nonlinearity produces a bias, and we have done a great deal of thinking about how one can correct for the bias. It is not a negligible effect, and it poses some special problems that we will have to solve.

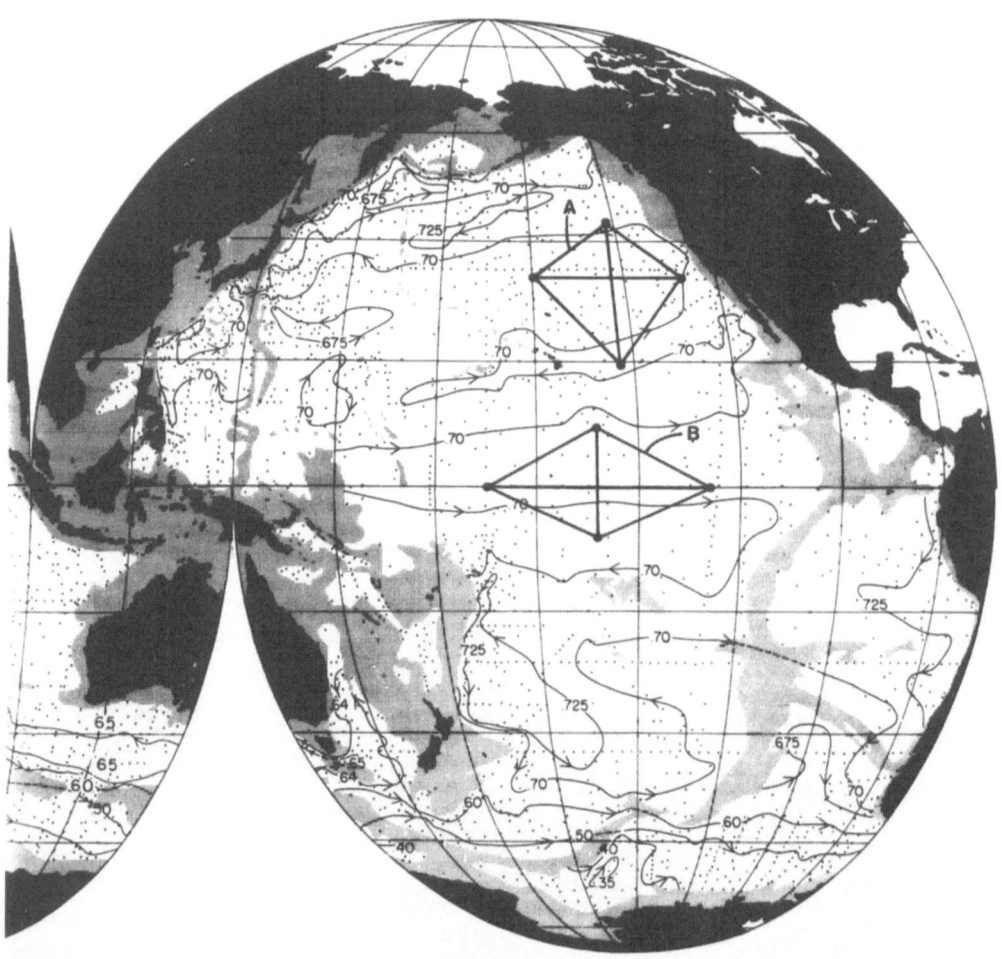

FIG. 9. Two proposed sites[2] for a large-scale tomography experiment in 1986.

References

1. W. Munk, "Acoustics and ocean dynamics," in *Oceanography: The Present and Future*, edited by P.G. Brewer (Springer-Verlag, New York, 1983), pp. 109-126.

2. W. Munk and C. Wunsch, "Observing the ocean in the 1990's," Philos. Trans. R. Soc. London **307**, 439-464 (1982).

3. B. Cornuelle, C. Wunsch, D. Behringer, T. Birdsall, M. Brown, R. Heinmiller, R. Knox, K. Metzger, W. Munk, J. Spiesberger, R. Spindel, D. Webb, and P. Worcester, "Tomographic maps of the ocean mesoscale -- 1: pure acoustics," J. Phys. Oceanogr. **15** (2), 133-152 (1985).

4. P.F. Worcester, "Remote sensing of the ocean using acoustic tomography," in *Advances in Remote Sensing Retrieval Methods*, edited by A. Deepak, H.E. Fleming, and M.T. Chahine (A. Deepak Publishing, Hampton, Virginia, 1985), pp. 1-11.

5. D. Behringer, T. Birdsall, M. Brown, B. Cornuelle, R. Heinmiller, R. Knox, K. Metzger, W. Munk, J. Spiesberger, R. Spindel, D. Webb, P. Worcester, and C. Wunsch, "A demonstration of ocean acoustic tomography," Nature **299** (5876), 121-125 (1982).

Astrophysical Jets

Roger D. Blandford
Professor of Theoretical Astrophysics
California Institute of Technology, Pasadena, California 91125

My subject is a topic in astrophysical fluid dynamics; specifically, the behavior of the supersonic jets that are observed to squirt out of the nuclei of active galaxies, compact stellar objects, and proto-stars. Some years ago, I had the good fortune to discuss this topic with Hans Liepmann, who quickly recognized that astrophysical jets pose some fascinating problems in fluid mechanics. However, he did not seem completely confident about the ability of astronomers to solve them (apparently some of my colleagues had been publishing papers on turbulence). Both of these insights turned out to be correct, as I hope to demonstrate in this talk.

Observations of extragalactic radio sources have recently been summarized and interpreted in several conference proceedings and review papers[1-6] and in one semi-popular article[7]. Most of the original technical literature can be traced through these references. Some of my figures in this paper are taken from a slide series[8] called "The Radio Universe" and are published here by permission of the National Radio Astronomy Observatory.

The fluid that I will discuss is a plasma, but it is one in which the effective mean free path is so much smaller than the scale of the system that we believe that the continuum approximation is excellent. This is partly because the Larmor radius is small, but also because plasma instabilities will make the bulk properties resemble those of a fluid in the usual sense. This is largely true in the solar wind, true in the interstellar medium, and also, we believe, true in the intergalactic medium. It is this belief that motivates the use of fluid mechanics in the particular problem of astrophysical jets. Actually, my subject turns out to be a bit more complicated than this, because these jets almost certainly pose a problem in magnetohydrodynamics, not just fluid mechanics. In this paper I will not emphasize the magnetic aspect. Nevertheless, I think the ultimate description must be magnetohydrodynamic.

I shall be primarily concerned with extra-galactic double radio sources. The first example of such a source, discovered in 1944, is known as Cygnus A. A modern map[9] is shown in Fig. 1. What is found is a pair of lobes of radio emission on either side of a distant galaxy. The regions of highest brightness are called "hot spots" and are found at the outer edges of strong sources; Cygnus A is one of the most powerful sources we know. This discovery of double radio sources was surprising, because the region of

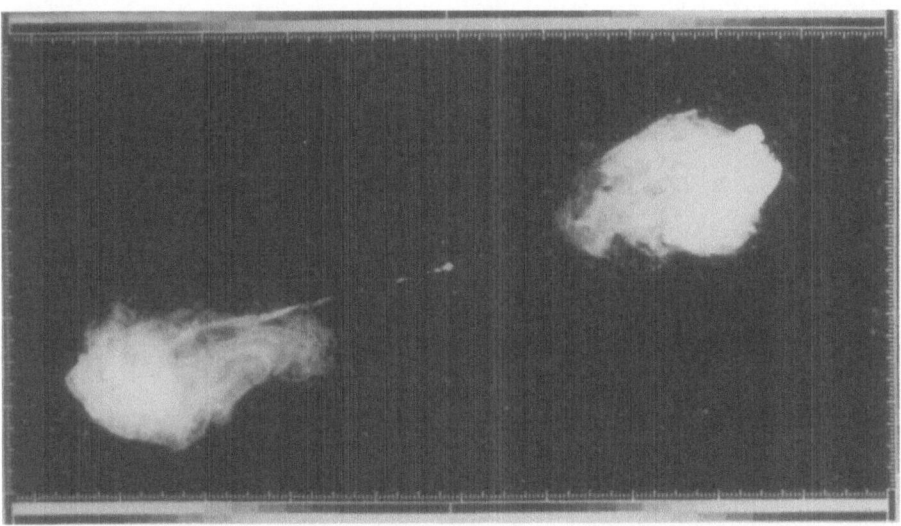

FIG. 1. A radio photograph[9] of the powerful double radio source Cygnus A (courtesy of NRAO). Clearly visible are the large radio lobes, the hot spots, one jet, and the central compact radio source identified with the nucleus of the associated galaxy.

radio emission was naturally expected to be located within the galaxy. However, it occurred well outside the optical image, and the question naturally arose as to how the underlying plasma came to be there. This is where jets come in.

Before I go on, I should say a little about diagnostics in this field and particularly about radio interferometers like the Very Large Array in New Mexico. A radio interferometer is a machine for recording a Fourier transform of the brightness of the sky and recreating the image numerically. This is the procedure that produces many of the maps in this paper. The emission that we detect in these sources is believed to be synchrotron radiation emitted by relativistic electrons spiraling in a magnetostatic field. By analyzing the brightness of the image we can learn three things. Firstly, we can estimate the pressure from the brightness and size of the source. Secondly, the polarization tells us the direction of the magnetic field; not the sign, but at least the direction. Thirdly, we get some partial density information from Faraday rotation measurements. However, we have very inferior diagnostics overall, and we have a much less well visualized flow than you are used to. In particular, we rarely have velocity information, and this lack is one of the big obstacles to progress in this field.

From maps like the map of Cygnus A, made in the early 1970's, it soon became apparent that what was going on was not, as was first thought, a massive explosion, but instead a continuous process. We infer that there are pipelines or channels or jets along which mass, momentum, and energy are transported in fluid form into the lobes. In Fig. 2 we can see, in a slightly better map[10] of a different source, a jet extending from the

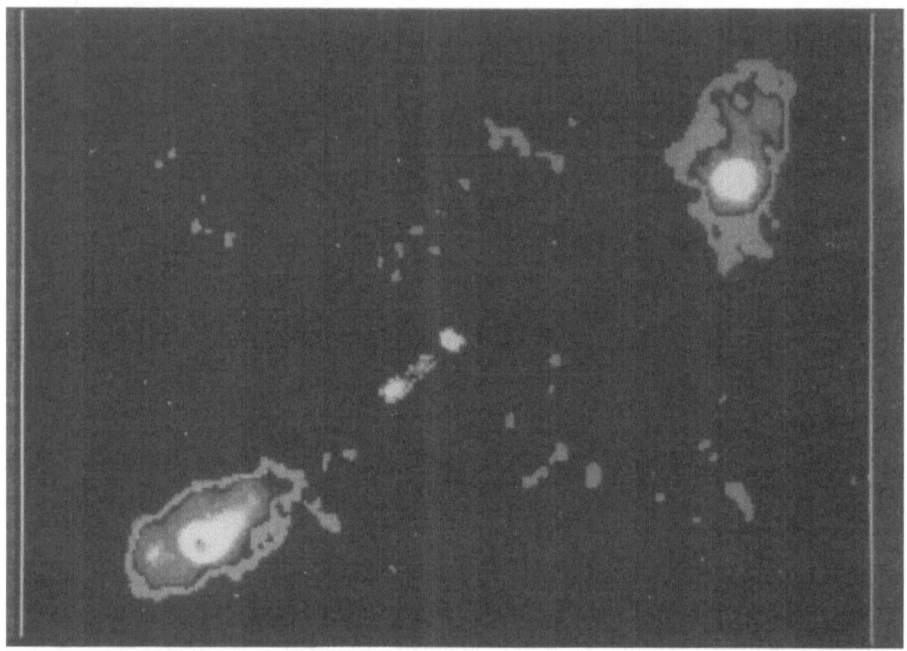

FIG. 2. A radio map[10] of 3C219, a powerful extragalactic radio source in which one jet is visible.

nucleus of the associated galaxy into the radial lobes. We now know of several hundred examples of these jets.

Linked interferometers like the Very Large Array are not the only instruments we have at our disposal. We can look in other parts of the electromagnetic spectrum, where several optical and X-ray jets are known. We can also look at radio frequencies with higher angular resolution, using the technique of Very Long Baseline Interferometry, in which the telescopes are not in the same place, but are distributed throughout the United States and Europe. By increasing the baseline we improve the resolution, so that we can look at details on a scale of light years in distant radio galaxies.

The results of Very Long Baseline Interferometry are exemplified by the view[11] in Fig. 3 of another radio source known as NGC 6251. First we see two large double lobes, 3 million light years across. The jets are transporting mass, momentum, and energy to the outer lobes from the nucleus of the associated galaxy. Next, we use VLBI to see what is going on right in the nucleus of the galaxy, on a scale of 3 light years. In these two figures we are looking at scales that span a range of a million to one. The lobes have been interpreted for a long time as jet flows at high Mach number ($M \geq 10$). The flow emanates from the nucleus of the galaxy and squirts out into the intergalactic medium, where it is finally brought to rest through a strong shock, or Mach disc. It is

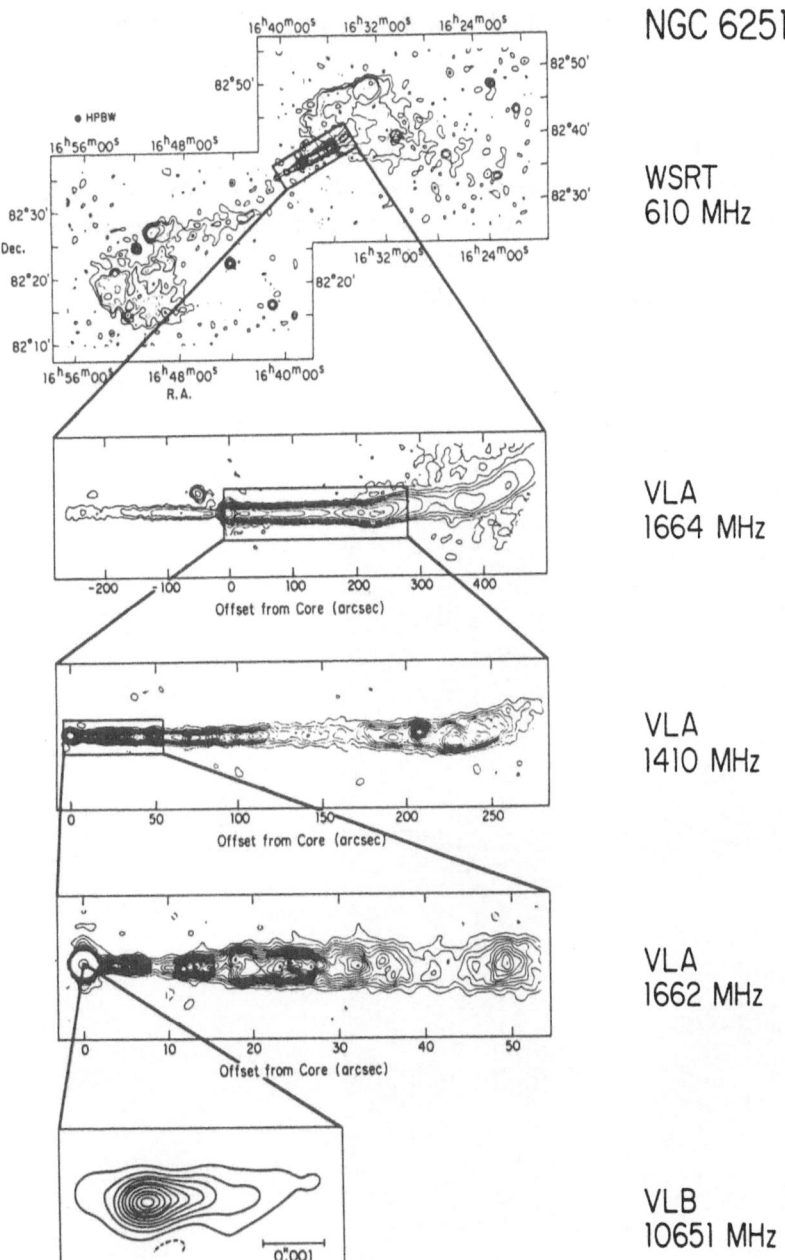

NGC 6251

WSRT
610 MHz

VLA
1664 MHz

VLA
1410 MHz

VLA
1662 MHz

VLB
10651 MHz

FIG. 3. A montage[11] of the intermediate-power radio source associated with the galaxy NGC 6251 (courtesy of NRAO). The jet directionality is maintained from length scales of light years (probed using VLBI) to length scales over a million times larger.

the strong shocks that are identified with the hot spots. After passage through the shock or Mach disk, there is a backflow that creates the cocoon, or the lobes of the radio galaxy.

The examples that I have shown so far are strong radio sources. Figure 4 is an example[12] of a weaker radio source associated with the galaxy M84. The galaxy is again located right in the center. The two jets do not terminate in strong shocks (hot spots) but appear to decay gradually with distance, petering out into a sort of plume rising buoyantly in the galactic gravitational field.

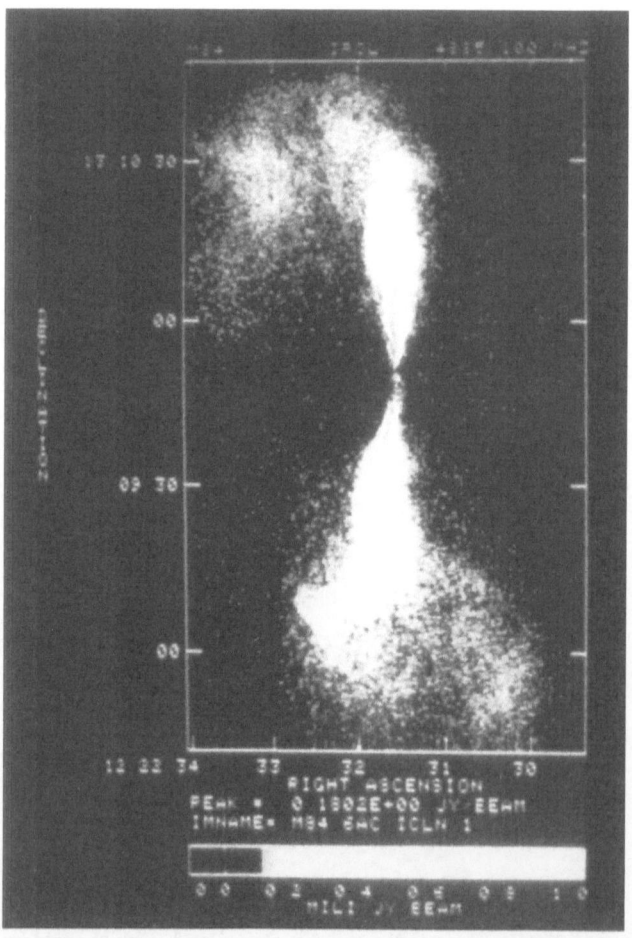

FIG. 4. The weak radio source[12] associated with the galaxy M84 (courtesy of NRAO). This is believed to be a subsonic or possibly transonic flow emerging from the central galaxy. Buoyancy in the galactic gravitational field may play a role in dictating the shape of the radio source.

One of the first questions that was asked when maps like these became available was: what is known in the laboratory about high-Mach-number, high-Reynolds-number jets? The answer is: not a lot. So perhaps we can turn the problem upside down and think about using these observations of extra-galactic radio sources as laboratories for studying flows at very high Mach numbers. Let me show a few examples with their associated interpretations. NGC 1265 in Fig. 5 is an example[13] of a radio trail. Again we see two jets emerging from the nucleus of the galaxy. This galaxy lies in a rich cluster and is moving hypersonically through the surrounding gas. The jets behave as if they were being swept back by the intergalactic medium. Indeed, although it is not apparent in this map, the radio emission extends much further. The jets appear to bend through an angle of more than 90° while retaining their integrity.

There are several interpretations of what might be going on in 3C449 (Fig. 6), which exhibits[14] several sharp bends. We might be looking at instabilities, about which I will have more to say. However, there is a crude reflection symmetry relating the two jets, and it could be that this galaxy is moving in dynamical orbit about its companions. If the jet has a very low velocity, then it can respond to acceleration of the source and may produce the observed shapes. Perhaps we can use these jets as tracers of the motion. By contrast, a radio map[15] of NGC 326 (Fig. 7) seems to exhibit inversion symmetry. This type of source has been interpreted not as a translational motion or orbital motion, but instead as a precessional motion of the source of the jet. The jet precesses about some axis fixed in space. In the past, presumably, the jet pointed in another direction.

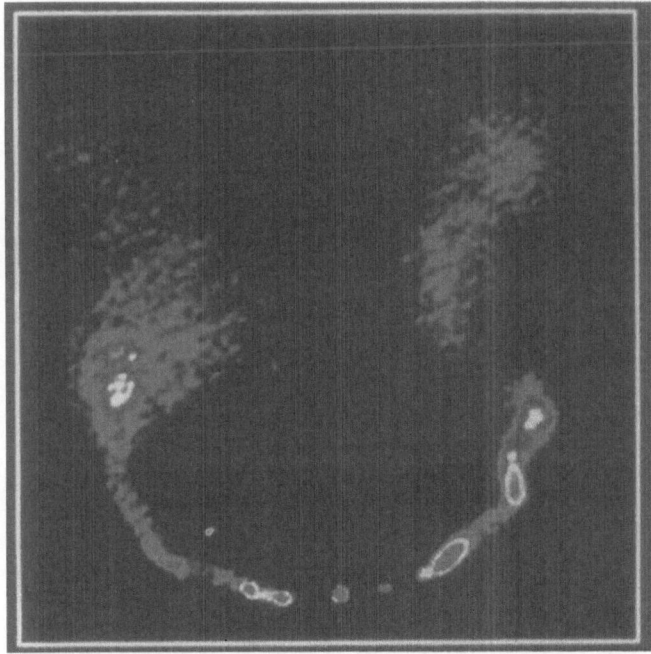

FIG. 5. The radio trail[13] associated with NGC 1265, which is believed to be an example of two anti-parallel jets propagating into a crossflow (courtesy of NRAO).

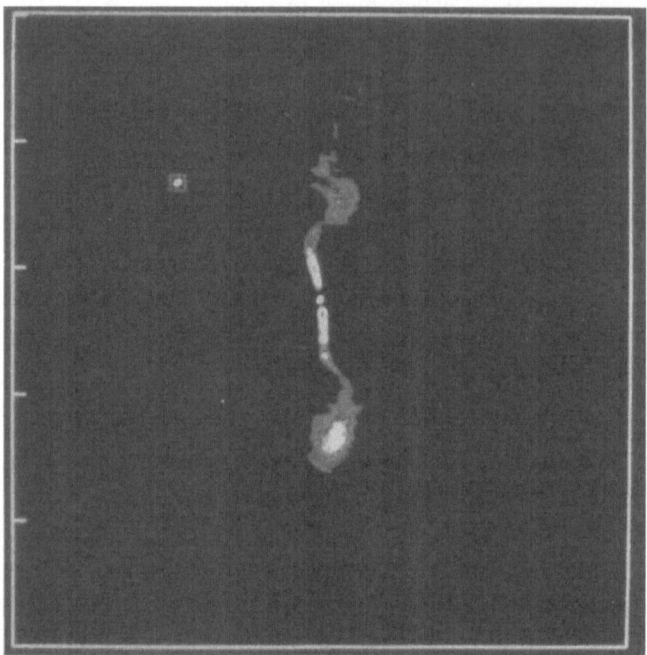

FIG. 6. 3C449, a source[14] exhibiting a crude reflection symmetry, possibly caused by a rapid acceleration of the parent galaxy (courtesy of NRAO).

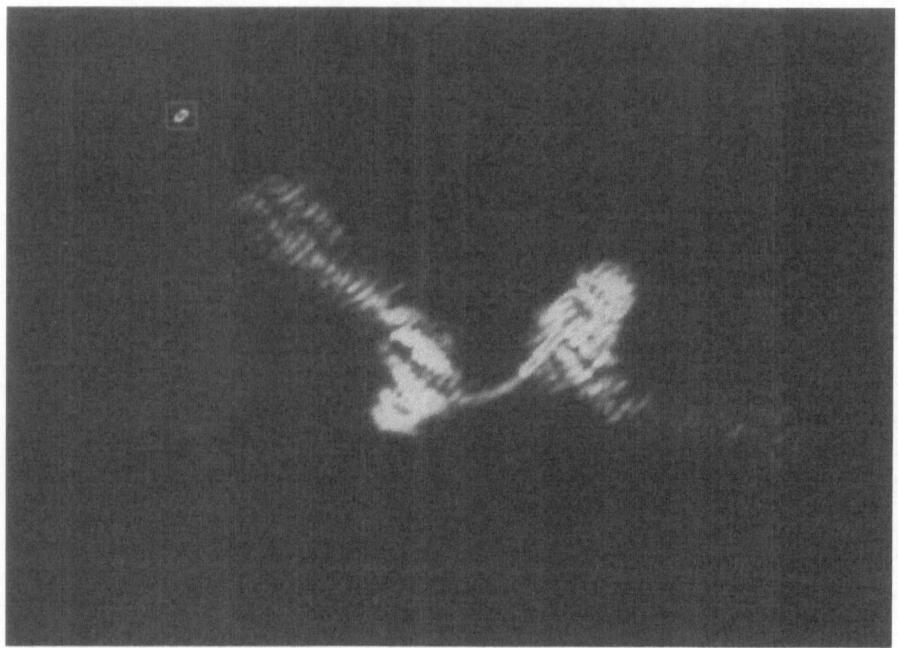

FIG. 7. The radio source[15] associated with the galaxy NGC 326 (courtesy of NRAO). The jets exhibit a crude inversion symmetry attributable to precession of their source.

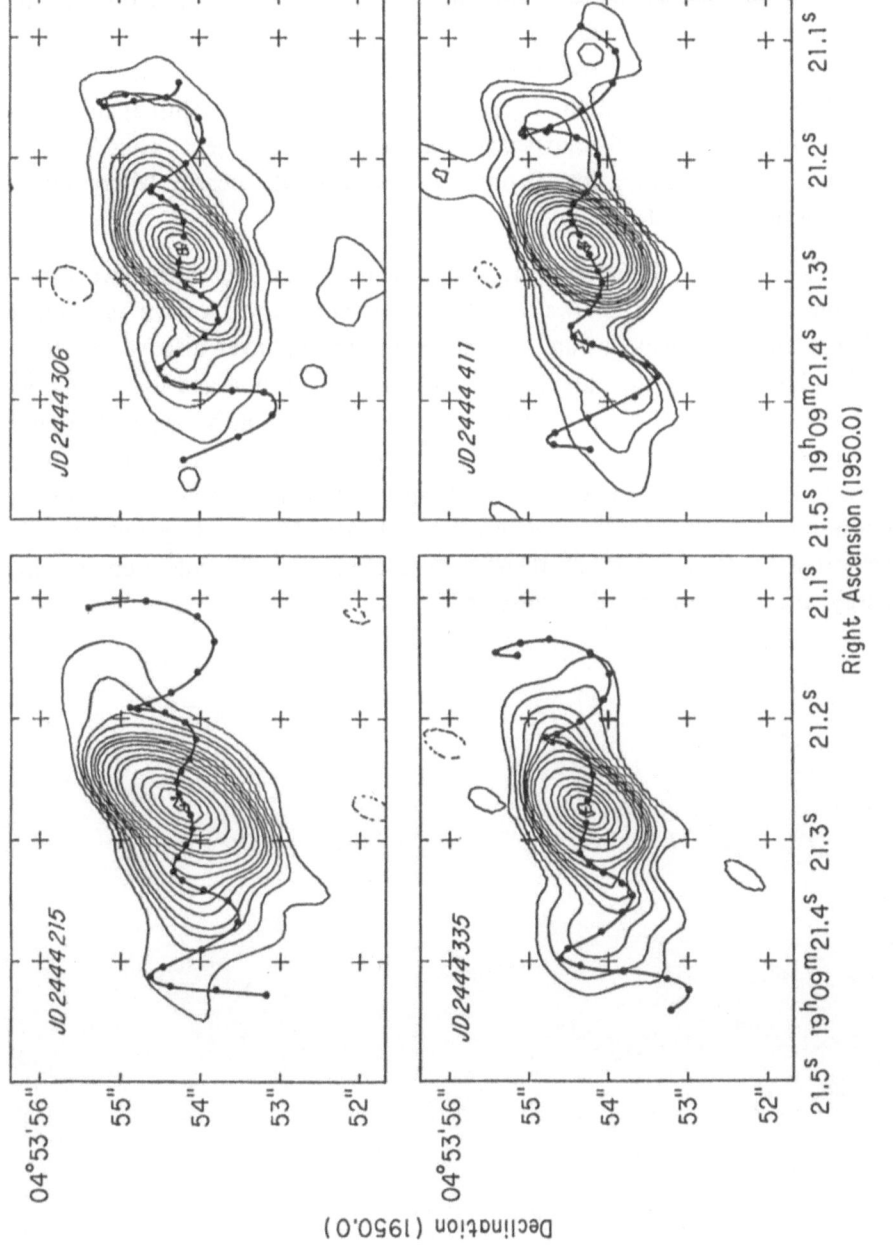

FIG. 8. Four radio maps[16], taken at different times, of the galactic object SS433. The precessional motion of the jets can be traced through the period of 164 days.

SS433 in Fig. 8 is another example[16] of a precessing jet. This source is known to be associated with a compact object within our galaxy, probably a neutron star or black hole in a binary system. Radio-emitting plasma squirting out in opposite directions from the source can be fitted to a particular ballistic trajectory for a precessing jet. Why am I so confident that the jet associated with this source is precessing? Because in this instance we can see the jet optically, and we are able to observe emission lines with their associated Doppler shifts. By analyzing these Doppler shifts as a function of time, we can infer that there are two outgoing jets moving at a quarter of the speed of light while precessing on a cone with a semi-apex angle of 20 degrees. I will show some of the data that support this interpretation. Figure 9 is a plot[6] of Doppler shift against time. The two sinusoids are what would be expected from two precessing jets. Thus SS433 is a source about which we know a lot. It provides a small-scale, nearby laboratory for studying more distant extra-galactic objects.

If astrophysical jets are interpreted as transonic or hypersonic flows, we expect shocks to be present. Figure 10 shows a famous external galaxy[17] known as M87. It lies in the Virgo cluster of galaxies and it contains the first jet ever discovered. If we take an optical photograph of the galaxy, on a scale of several thousand light years, we can see a jet emerging from the center of the galaxy. We see it in X-rays and in radio waves as well. Now this jet has several features called knots. We suspect that these knots are instabilities or shock fronts associated with outflowing gas in the jet. Figure 11 is a high-resolution radio map[18] of the brightest of these knots, which indeed looks like a shock front. Behind the shock there is dissipation of bulk kinetic energy into internal energy, and synchrotron radio emission.

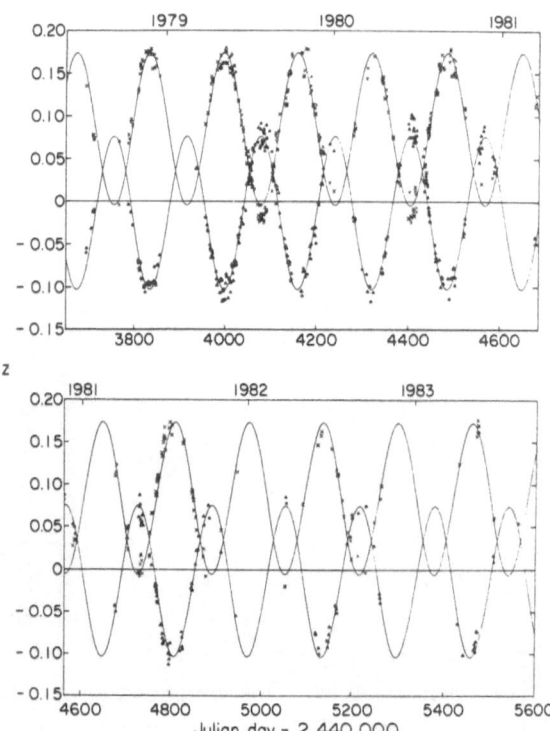

FIG. 9. Doppler shifts[6] of the emitting gas in SS433 over the period 1978-1983. The data are well fitted by a precessing-jet model. The emission lines are created much closer to the origin than the radio components shown in Fig. 8.

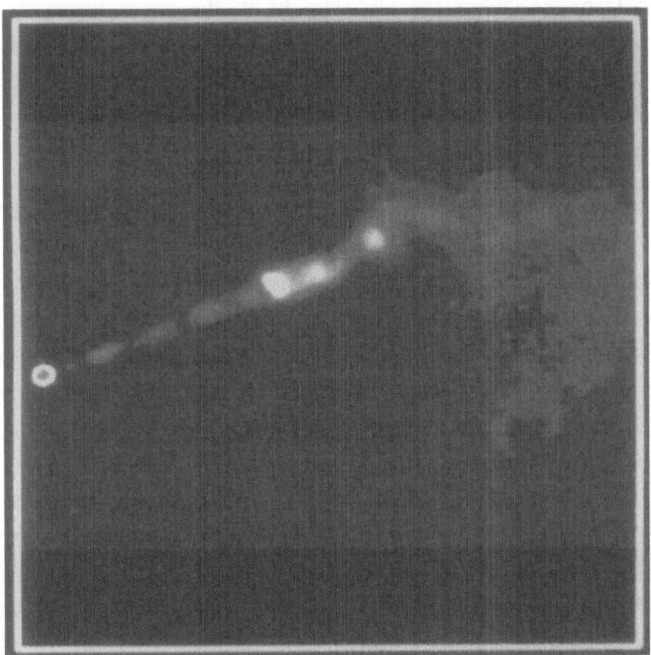

FIG. 10. The radio jet[17] emerging from galaxy M87 in the Virgo cluster (courtesy of NRAO). This jet is seen at optical and X-ray wavelengths as well.

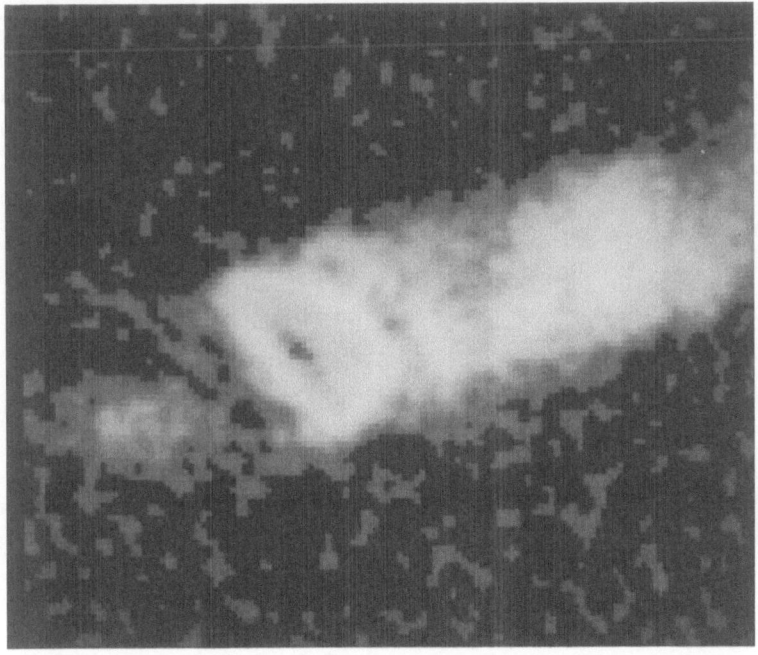

FIG. 11. A higher resolution map[18] of the brightest knot in the M87 jet. This may be a shock front.

In addition to these internal shocks, there are also shocks in the surrounding gas. Hercules A provides a good example[19] (Fig. 12). Intermittency in the jet may be driving strong compression waves or weak shock waves into the surrounding gas, highlighting the radio emission. The features here can certainly be interpreted in these terms, although there are alternative possibilities.

If all of this complexity is not enough, there is one more feature that is important, particularly in strong sources. As some of you may have noticed, the jets in the strong sources appear to be one-sided. We see the radio emission only on one side of the nucleus, not on the other side. One explanation that I favor is that these jets are moving at nearly the speed of light, and they beam their radio emission in the forward direction; a jet that is coming toward us appears substantially brighter than a jet that is receding. We have some circumstantial evidence for this interpretation in Fig. 13, which shows a montage[20] of four successive VLBI maps of the quasar 3C273, the first quasar discovered here at Caltech. We see features apparently moving outwards from the origin of the jet, with the displacement increasing from 62 light years to 87 light years during the period from July, 1977 to July, 1980. The motion is faster than the speed of light. This observation is not a refutation of the special theory of relativity. Instead, we believe that it is a kinematical illusion that can be understood in terms of light travel-time effects. I won't go into that explanation, which is an exercise in freshman physics, beyond remarking that it requires the jet fluid to move at relativistic speed. Not only do we have to understand hypersonic high-Reynolds-number flow; it has to be relativistic as well.

I have touched briefly on a lot of material. Let me try to summarize to this point. Astrophysical jets are common and surprisingly persistent. Jets at high Mach number in the powerful radio sources are harder to see. What this means, in crude terms, is that they are less dissipative. There is less conversion of bulk kinetic energy into internal

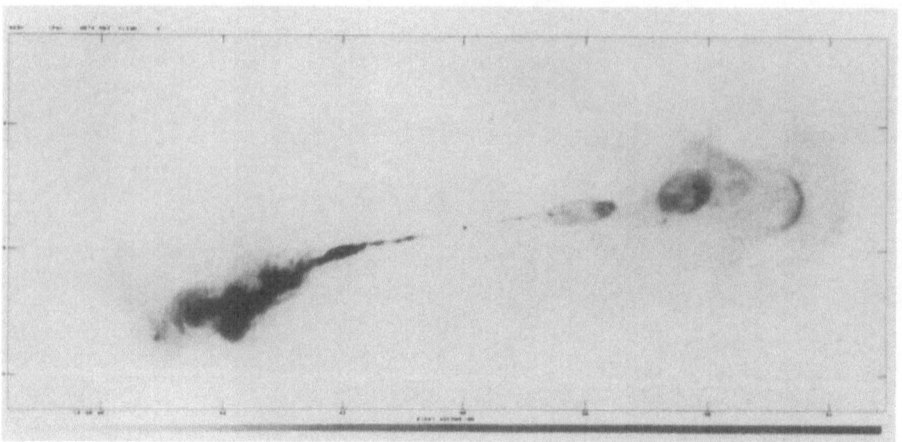

FIG. 12. A negative image[19] of the radio source known as Hercules A (courtesy of NRAO). Note the bright circular shells, which may be associated with shock fronts.

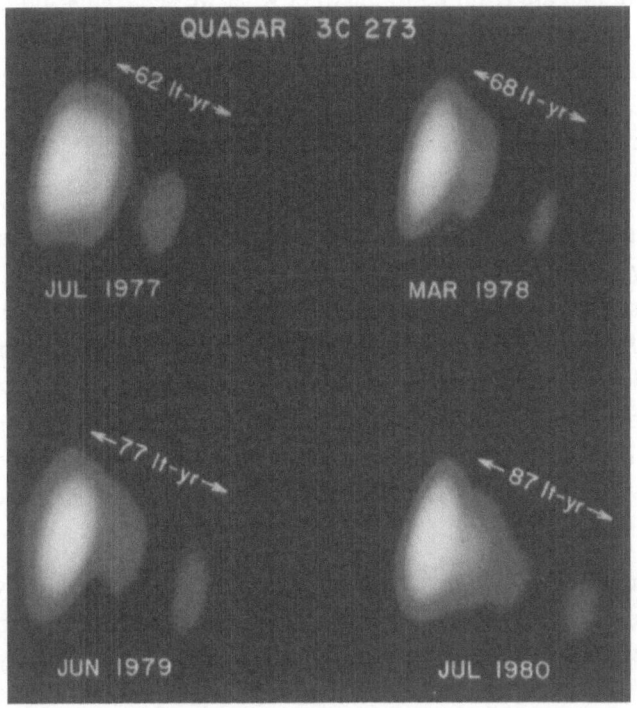

FIG. 13. A montage[20] of the small-scale radio jet in the quasar 3C273 as seen at four successive epochs. The bright features appear to be moving faster than the speed of light.

energy. They are also frequently terminated by strong shocks, making the hot spots that I alluded to earlier. The power levels are so high that we have to be very grateful for the scaling laws that operate in fluid mechanics. By contrast, jets in the weak sources appear to be transonic or subsonic. They may start out at low Mach number and then eventually degenerate into plumes and bubbles. It appears to be characteristic of the lower-Mach-number jets that they are much more dissipative, as you might indeed expect. Because they are more dissipative, they are easier to study. Another reason, of course, is that they are nearer to us.

Let me mention some numbers designed to give a quantitative feeling for the scale of these sources. A strong source like Cygnus A has a power of approximately 10^{38} watts and a thrust of 10^{30} Newtons. The magnetic fields within the radio lobes are comparatively weak by terrestrial standards ($< \sim 10 \, nT$) but the total energy involved ($> \sim 10^{52} \, J$) is the rest-mass equivalent of $\sim 10^5$ stars. The largest radio sources, some 10 million light years across, contain up to a hundred times more energy.

How can we learn more about these fascinating objects? Four lines of attack are being followed. The first is to make simple numerical estimates of the levels of energy, power, magnetic flux, and so on, that are required to interpret what we see. This sort of arithmetic is reasonably straightforward and correspondingly inaccurate. The second method is an appeal to experiment. The third method, one that is just coming of age and will develop rapidly in the future, is use of computer simulations. Finally, we can perform analytical calculations that model equilibrium flows and explore their stability properties. I have time to give you only the flavor of these four approaches.

One fundamental question to ask about astrophysical jets is: what is the ratio of the density in the jet to the density of the surrounding gas? One way to inquire about this is to perform experiments. Drs. Kieffer and Sturtevant have begun some work[21] in this direction. In Fig. 14a a jet of light gas, helium, is squirted into a heavier gas, nitrogen. We see a bubble forming and the jet maintaining its integrity, with some back flow around the sides. When nitrogen flows into nitrogen, as in Fig. 14b, we see a strong vortex being formed as the jet breaks up. A heavy freon jet in Fig. 14c propagates into nitrogen essentially ballistically. Features that are exhibited in these photographs have morphological counterparts in the radio maps.

Next consider some numerical calculations[22], in this case carried out on a Cray computer by Norman and Winkler. In Fig. 15 we see a propagating axisymmetric Mach 12 jet of a perfect gas, with different values for the density ratio. When the jet density is much larger than the density of the external medium, the jet propagates more or less ballistically. A light jet terminates in strong shock fronts, followed by backflow which creates a cocoon-like structure. The whole pattern advances through the surrounding medium and is bounded by a strong shock wave. I should warn you that there are some difficulties in associating features in computed jets with observations, because the computations use a very simple gas-dynamic model without dissipation, whereas what

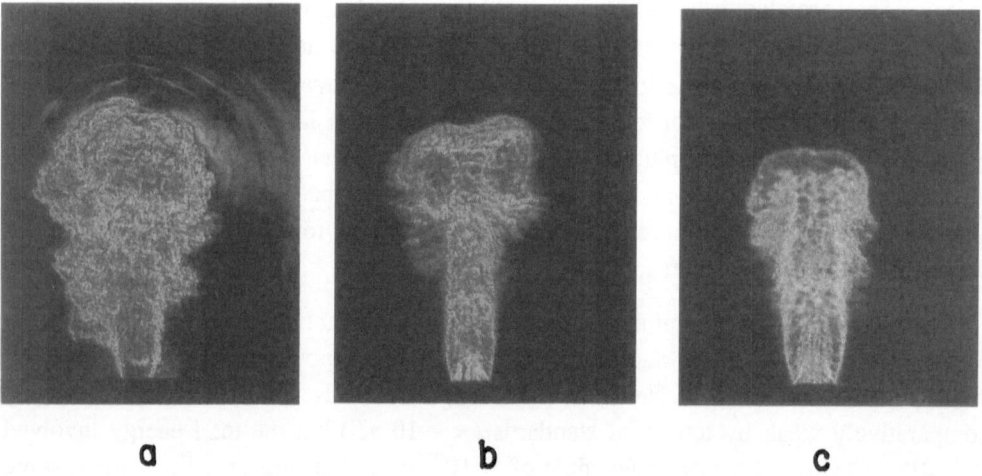

a   b   c

FIG. 14. Laboratory experiments on starting jets[21]. (a) helium into nitrogen; (b) nitrogen into nitrogen; (c) R22 (chlorodifluoromethane) into nitrogen (courtesy of B. Sturtevant).

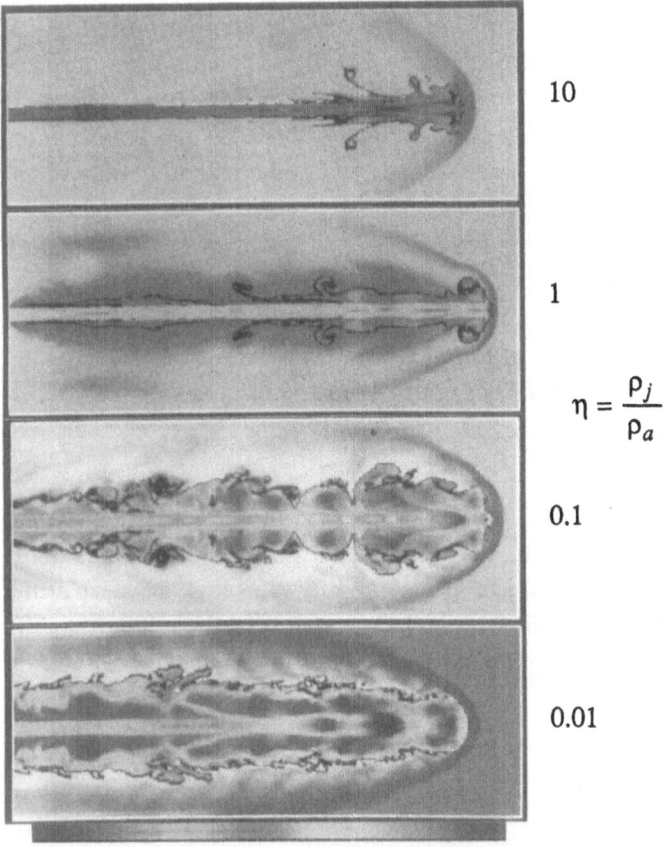

10

1

$$\eta = \frac{\rho_j}{\rho_a}$$

0.1

0.01

FIG. 15. Numerical simulations[22] of Mach 12 jets carried out on a Cray supercomputer (courtesy of Los Alamos Science). The density field is shown for different values of the ratio of initial jet density to ambient density.

we are looking at in the radio maps is essentially dissipation, and even that is rather imperfectly diagnosed.

We have known for a long time that the jets are unstable. In fact, the chief surprise of this subject for many people has been that the jets persist for such a long distance, whereas we might expect Kelvin-Helmholtz and other instabilities to disrupt them. There has been a lot of analytical work on understanding Kelvin-Helmholtz instabilities. Most of it has been confined to the linear regime, which means small perturbations that cannot be seen by radio telescopes. Rather than reproduce the dispersion relations, I will try to summarize qualitatively what appears to be going on. If we confine our attention to the axisymmetric modes, then there are two basic types. One we call the ordinary modes; these have no nodes inside the channel. The other modes are the so-called reflecting modes, which involve waves propagating back and forth across the channel. A speculation that appears to be borne out by numerical computation is that the ordinary modes are much more destructive than the reflecting modes. When we plot the density ratio η as ordinate against the Mach number M as abscissa, as in Fig. 16, then the four cases in Fig. 15 are the ones with $M = 12$. If we make a linear analysis of flow conditions in a jet of given density ratio at Mach number M, then in the upper left part of the diagram it turns out that the ordinary mode is more disruptive. Essentially what happens is that a mixing layer spreads across the jet and decelerates it, and this eventually leads to termination after something like 20 to 50 jet diameters. Indeed, these computations are claimed to have considerable similarity to laboratory experiments. The other case is at a lower density ratio. Here we believe that the less disruptive reflecting modes are important. There is much less of a mixing layer, and we get a lot of criss-crossing internal shocks inside such jets. Those are readily interpreted as knots and bright features, but they turn out not to be particularly disruptive. When we have a very high Mach number and a very low density ratio we see cocoons like those in Fig. 15.

I hope I have given you some idea of the problems posed by astrophysical jets and the techniques that are being brought to bear on them. To a physicist, the most interesting question is: how are these jets made? Many people believe that the production of jets is associated with gas flow around a massive black hole lurking in the nuclei in most of these galaxies. In one of the most ambitious numerical simulations yet

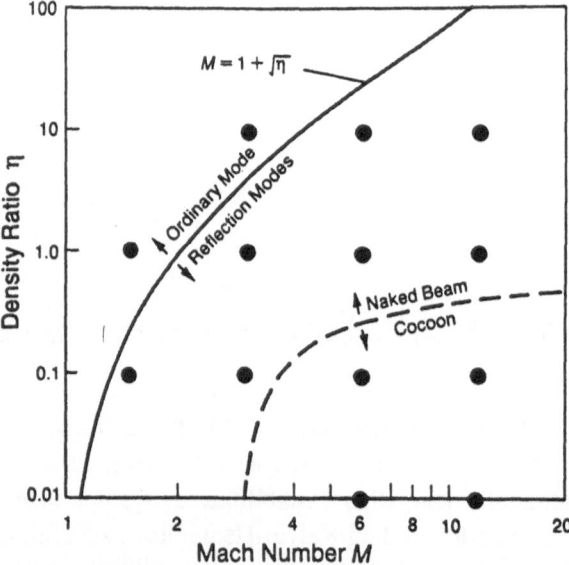

FIG. 16. Results of simulations[22] of numerical jets (courtesy of Los Alamos Science). The jets that show ordinary-mode instability are disrupted more violently than those showing reflection modes. The powerful jets associated with extragalactic radio sources are believed to be highly supersonic and to have a low density ratio, and therefore lie in the bottom right-hand corner of the diagram.

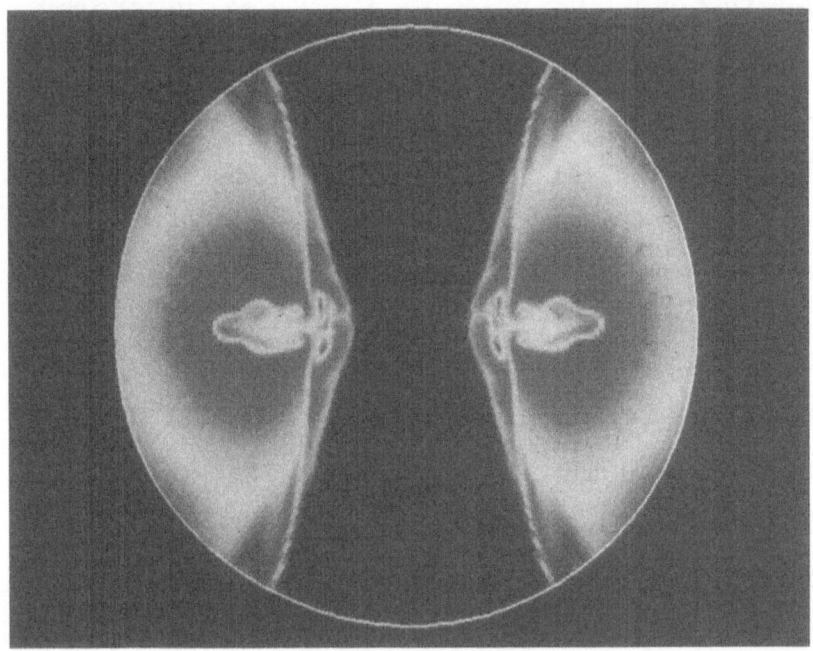

FIG. 17. One frame from a computer-generated movie[23] showing accretion of gas with angular momentum onto a black hole. Note the formation of funnels, which may be related to the production of jets (courtesy of J. Hawley).

attempted, John Hawley has made a computer movie[23] of flow about a black hole (Fig. 17). Using such simulations, we are slowly developing some intuition about the behavior of gas flowing in these exotic environments. However, impressive as these simulations are, I share Julian Cole's implied prejudice yesterday that compressible flows are not just an exercise in computing. A judicious combination of observation, analysis, experiment, and numerical work will be necessary before we properly understand these fascinating cosmic objects. I promise to come back to Hans Liepmann's 80th birthday celebration to report on progress.

References

1. D.S. Heeschen and C.M. Wade (eds.), IAU Symp. No. 97, *Extraglactic Radio Sources* (Reidel, Dordrecht, 1982).

2. M.C. Begelman, R.D. Blandford, and M.J. Rees, Rev. Mod. Phys. **56**, 255-351 (1984).

3. R. Fanti, K. Kellermann, and G. Setti (eds.), IAU Symp. No. 110, *VLBI and Compact Radio Sources* (Reidel, Dordrecht, 1984).

4. A.H. Bridle and R.A. Perley, Ann. Rev. Astron. Astrophys. **22**, 319-358 (1984).

5. M.J. Rees, Ann. Rev. Astron. Astrophys. **22**, 471-506 (1984).

6. B. Margon, Ann. Rev. Astron. Astrophys. **22**, 507-536 (1984).

7. R.D. Blandford, M.C. Begleman, and M.J. Rees, Sci. American **May 1982**, 124-142.

8. Slide series "The Radio Universe," National Radio Astronomy Observatory, Tucson, Arizona 85705.

9. Ref. 8, slide 07, "Cygnus A radio galaxy (VLA)."

10. A.H. Bridle, R.A. Perley, and R.N. Henriksen, Astronom. J. **92**, 534-545 (1986).

11. Ref 8, slide 14, "NGC 6251 radio galaxy—zoom" (see also Ref. 4).

12. Ref. 8, slide 30, "Radio galaxy M84 (VLA)."

13. Ref. 8, slide 29, "Radio galaxy NGC 1265 (VLA)."

14. Ref. 8, slide 26, "Radio galaxy 3C449 (VLA)."

15. Ref. 8, slide 24, "Radio galaxy NGC 326 (VLA)."

16. R.M. Hjellming and K.J. Johnston, in *Extragalactic Radio Sources*, edited by D.S. Heeschen and C.M. Wade (Reidel, Dordrecht, 1982), pp. 197-203.

17. Ref. 8, slide 35, "Jet in M87 (VLA)."

18. J.A. Biretta, F.N. Owen, and P.E. Hardee, Astrophys. J. **274**, No. 1, Part 2, L27-L30 (1983).

19. Ref. 8, slide 10, "Hercules A radio galaxy (VLA)."

20. T.J. Pearson, S.C. Unwin, M.H. Cohen, R.P. Linfeld, A.C.S. Readhead, G.A. Seieldstad, R.S. Simon, and R.C. Walker, Nature **290**, 365-368 (1981).

21. B.S. Sturtevant (private communication).

22. M.L. Norman and K.-H. A. Winkler, Los Alamos Science, No. 12, **Spring-Summer 1985**, 38-71.

23. J.F. Hawley and L.L. Smarr, "Evolution of a thick-disk instability," 16 mm, color, 6 min. (1984) (private communication).

Natural and Artificial Flying Machines

Paul B. MacCready

AeroVironment, Inc., Monrovia, California 91016

Summary

The advent of fossil-fuel engines has provided aeronautical engineers with a ten-fold to hundred-fold increase in power-to-gross-weight ratios over the ratios available for biologically-powered flight creations, such as birds and human-powered aircraft. The tremendous achievements of engine-powered aircraft over the past eight decades have tended to obscure the fact that numerous flight problems had already been elegantly solved by birds many tens of millions of years ago. Recent projects in human-powered aircraft, in bird aerodynamics, and in the development of a flying replica of a 11-m span pterodactyl have introduced us to the bird-airplane interface. The result has been an increasing respect for "Mother Nature the Engineer," who derived efficient evolutionary solutions for all of the factors involved in biological flight. Engineers and scientists also have much to learn from nature regarding aeroelasticity as a factor in tailoring structures to the varied demands of flight, including active-control technology, boundary-layer control, and navigation.

I. Introduction

Early aviation derived its inspiration primarily from the role model of birds, and some early flight attempts even involved feather substitutes and bird shapes. After the successes of Cayley, Lilienthal, and the Wrights, the development of gasoline power plants, and contributions to the theoretical foundations of the field by Lanchester and Prandtl, man's aviation constructions raced far beyond those of birds, and the role model became virtually forgotten. We all observe birds and admire their grace, beauty, and freedom, but their role in aviation has been relegated more to worries about avoiding ingesting them into jet engines or cleaning their signatures off wings than considering them as creatures offering useful insights to designers.

Basic research about birds and their evolution is increasing, but if we recognize at all a connection to aircraft, the connection is likely to be only the after-the-fact realization that many modern design solutions could have been anticipated by observing how nature has been doing it for millions of years. Use of nature's designs to help us solve new aeronautical problems is rare. Perhaps the appreciation for evolution as a master designer of aeronautical form and function best suits the sailplane field. Sailplanes, like soaring birds, must be very efficient and must be operated efficiently to utilize nature's invisible lift, and so sailplane pilots and designers still observe birds carefully and learn something from them.

Considerable literature exists about natural flight. As general reviews, I recommend the Symposium on Flying and Swimming in Nature[1] and the papers by Kuethe[2] and McMasters[3], each of which has helpful reference lists. The latter two also make many comparisons of natural and artificial flying devices. The focus of the present paper is on selected items not covered in these references, although some overlap is unavoidable.

Circumstances have involved me with the interface between natural flight (birds, pterosaurs, insects, etc.) and artificial flight (airplanes). My explorations have been stimulated especially through the subject of human-powered flight, wherein natural muscle is integrated with artificial structure and mechanisms. These explorations have emphasized low power loading, a focus enforced by the inefficiency of muscle as compared with the internal combustion engine. The explorations have thus also emphasized aerodynamic efficiency and light-weight structures, which permit flight with low power loadings.

The outcome from all this for me has been a growing realization that Mother Nature is a fantastic aeronautical engineer and has been so for many tens of millions of years. Nature utilizes evolution to develop solutions for filling ecological niches. There are continuing variations of creatures and continuing survival pressures. Statistically, only winners survive to leave progeny. In contrast to scientific developments in civilization, mistakes tend not to be respected as learning experiences, and second chances are rare. Incompleted "projects" in nature cannot be rescued by a sponsor picking up overrun costs. As a result, natural engineering is pragmatic, complete, and effective. Birds have solved myriads of problems in aerodynamics and structures, including problems scientists have not even recognized yet. Identifying and investigating these solutions represent a fertile research opportunity.

II. Some background factors and perspectives

A number of events and projects have served to stimulate my enthusiasm for nature's engineering of flight devices. A brief review here of these events and projects will set the stage for the comparisons that follow of natural-vs-artificial aeronautical devices, with the review illuminating some key points.

In the late 1930's, my hobby of model-airplane flying introduced me to the comparison of man's constructions with birds, and to an appreciation for the effectiveness of birds in locating and using thermals. From 1945 to 1956, a commitment to sailplanes and soaring as scientific research topics further fanned my interest in and awe of the flight of soaring birds. A paper by Woodcock[4], titled "Soaring Over the Open Ocean," made a deep impression on me because it seemed to be an ideal scientific experiment, one which had significance and yet could be conducted without any special equipment.

Woodcock watched soaring birds during a long ocean voyage; saw whether they soared in circles, straight lines, or did not soar at all; noted the wind speed and temperature difference between air and water; and found that the atmospheric flow patterns indicated by the soaring techniques used for various winds and stability conditions were analogous to the patterns of Benard cells in liquids, a problem which has long been studied in laboratories and for which quantitative theory is well developed. Thus, motions on scales of millimeters in laboratory liquids (involving molecular transfer) can be related to motions at a millionfold increase in scale in the atmosphere (involving turbulent eddy transfer), with bird observations providing the key data link.

Somehow, advancing science by watching birds soaring seemed to me to be an elegant research technique. In 1976, on a rare family vacation driving across the U.S., I realized that certain simple observations of birds in circling flight could provide valuable information on their aerodynamic capabilities. In fact, even the average lift coefficient of the airfoil in flight could be inferred. The vacationing study had two fruitful outcomes, beyond an initial informal paper[5]. For one, my comparison of flight characteristics between birds of different species and hang gliders and sailplanes served as the catalyst for the ideas behind the development of the Gossamer Condor[6-10]. The other outcome was a more careful investigation in 1980 and 1982 into the flight characteristics of frigate birds, possibly the best of all natural soarers[11]. This latter study suggested that a) frigate birds may sometimes operate at a surprisingly high lift coefficient, higher than we would have expected at the operative Reynolds number, and b) the birds significantly alter the details of their thermalling flight mode with meteorological conditions, as do sailplane pilots.

Our human-powered airplane projects (Gossamer Condor, 1976-1977; Gossamer Albatross, 1978-1979; Bionic Bat, 1984-1985) focused our attention on the interrelation between birds and airplanes. Henry Kremer's prize challenge was to use human power to fly. A human has a low power-to-weight ratio, but one probably not greatly different from that of a soaring bird. In any case, the ratios of power to gross weight for these biologically-powered vehicles are about two orders of magnitude less than the ratios for aircraft powered by internal-combustion engines. The low power-to-weight ratio is compatible with flight with a low wing loading and hence low speed.

Attention to low-speed flight stimulates attention to the effects of atmospheric turbulence on efficiency and controllability. Performance of the Gossamer aircraft deteriorated rapidly with increasing turbulence. At a flight speed of only 5 or 6 m/s, a gentle local upcurrent or downcurrent of 0.5 m/s means a local angle-of-attack change of about 5 degrees, with consequent adverse effects on induced drag and parasite drag. The effects on stability can be even more significant, as the effective angles of attack of surfaces can exceed stall limits and the control limits of ailerons. Extreme care had to be exercised in flying our solar-powered Solar Challenger at about 10 m/s in turbulence near the ground. Similarly, operation of hang gliders near the ground, at comparable speeds, emphasizes control limitations in turbulence. Birds have the brain and muscle to articulate their wings as dictated by the local airflow, and hence can fly without problems in turbulent conditions that trouble these piloted aircraft.

Finally, my growing respect for and envy of nature as a designer of aeronautical creatures got another boost recently as a consequence of my starting on a project to create a flying replica of a giant (11-m wingspan) pterodactyl. Not only did the size go well beyond the limiting size of natural flying creatures as inferred from extrapolation of standard scaling laws, but the tailless flier probably had a wing that was unstable in pitch, and so the pterodactyl must have used some means of active control (wing sweep?) to provide effective stability. A search for literature on bird pitch stability and controllability was generally unfruitful. Birds such as the albatross, the gannett, and even the sea gull in smooth slow gliding employ essentially no tails. Their active control systems deserve study.

Starting with this background, it seemed reasonable to me to explore broadly just what flight-related features nature may have developed prior to civilization's technological aeronautical achievements.

III. Overview of natural vs artificial flight and fliers

There are several major areas where birds (or other natural fliers) cannot be expected to be directly analogous to airplanes. One is in transonic and supersonic flight, which is certainly man's prerogative alone (natural flight evolution was never concerned with aerodynamic effects at the speed of sound). Another area is the power system. The high energy density obtainable with fossil fuel, and especially the high power-to-weight ratio, let airplanes achieve speeds, altitudes, and load-carrying ability that are beyond consideration for birds. For the most part, the best direct correlations between natural and artificial flight should arise from the larger natural fliers vs the smaller and slower airplanes.

An obvious difference between birds and aircraft is that propulsion in birds comes from flapping wings, while in airplanes it comes from rotating machinery (propellers, either exterior to or integral with the engine). But for each propulsion method the mechanical-aerodynamic propulsion efficiency during normal flight is usually within ± 10 percent of 85 percent. Thus the one method does not offer any great advantage over the other for propelling the vehicle.

The bird offers great features of versatility. For example, the loon is effective in flying through the air, walking on the ground, and operating on and under water. No doubt a manned airplane could be constructed to do the same, but the undertaking would be formidable.

A bird's versatility is to some extent associated with the use of parts for multiple functions. For example, a bird's wings are used for propulsion, lift, and stability and control, with variable geometry for different flight modes; they also serve for ornamentation, and for insulation when retracted. With an airplane the function of each part tends to be more specialized. For the ultimate in efficiency; e.g., a sailplane with a best glide ratio exceeding 60:1, the separation of function is distinct. The wing handles lift efficiently (and roll control), the fuselage handles the payload (and supports the landing gear and tail), and the tail provides yaw and pitch stability and control. To use the pitch-control device to contribute to lift (a canard), or to ask the wing to handle the stabilizer-elevator task (a flying wing), compromises vehicle efficiency even though it may offer benefits in other areas. For an airplane, the total flight system can be modified to permit emphasis on efficiency where it is needed. A long runway permits an airliner's design to emphasize cruise efficiency; if a bird-like takeoff from a tree or from unimproved ground were required, the vehicle would be more like a helicopter or Harrier jet, with much lower cruise efficiency and payload capability.

Nature has achieved full flight by at least four separate routes: birds, mammals (bats), reptiles (pterosaurs), and insects. For more limited flight we can even include flying fish and gliding animals and seeds. When filling a particular ecological niche involving, say, a flying animal with a wing span of about one meter, the rules of conservation of energy and momentum, the realities of viscous flow phenomena, and the limits of biological power and biological structure dictate that nature finds rather similar solutions no matter what the starting point. Figure 1 illustrates the different skeletal solutions by birds, mammals, and bats to the problem of producing a wing[12]. Where birds and insects overlap, as with a hummingbird and a hawk moth (hummingbird moth), the appearance and function, both for flight and for feeding on the nectar of flowers, are remarkably similar, although the inner structural details are quite different.

Human engineers face the same aerodynamic realities involving energy, momentum, and viscosity as do flying animals, but the engineer can utilize structure strength-to-weight and propulsion power-to-weight ratios that are many times those available to biological systems.

IV. Birds vs airplanes

A. Long flights

Aircraft with air-breathing engines have stayed aloft 84 hours and covered almost half the circumference of the earth without aerial refueling. With refueling, the duration rises to 64 days, and the distance to more than once around the world. Birds hold their own in the duration and distance categories:

1. The sooty tern can stay aloft for years at a time. Of course it uses aerial refueling, primarily by snatching food (fish and squid) from the ocean surface without alighting.

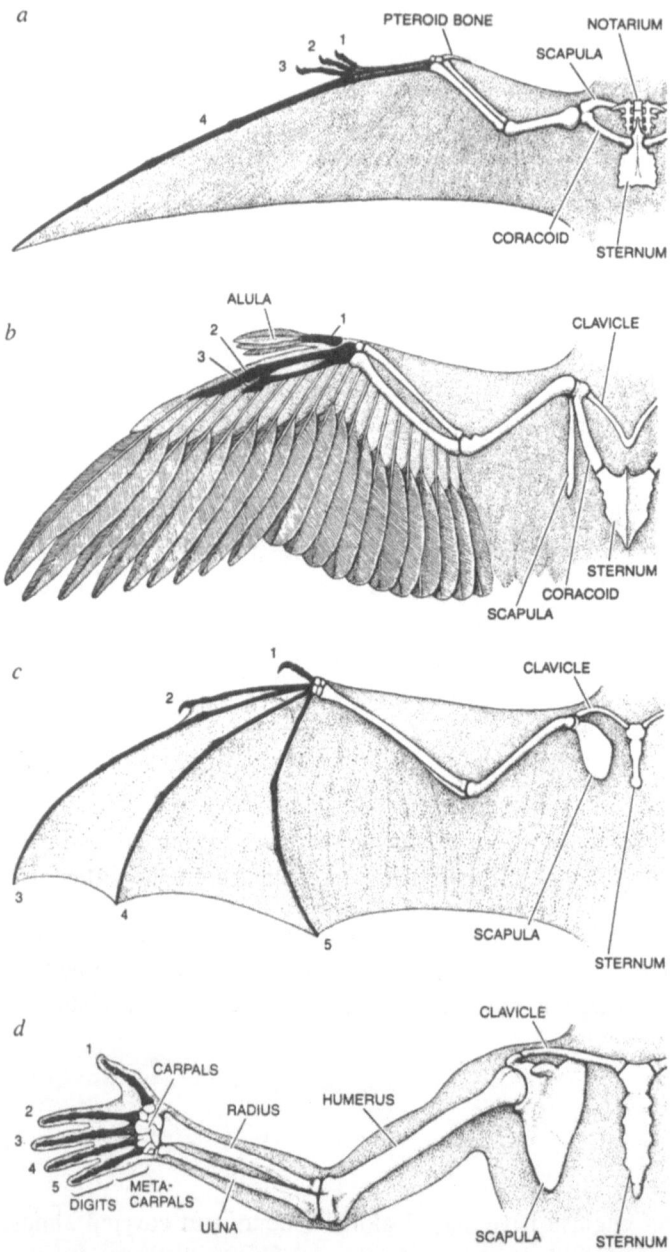

WINGS of a pterosaur (*a*), a bird (*b*) and a bat (*c*) are evolutionary variations on a forelimb that was suitable for an earthbound animal that walked on all fours. The variations are distinctive: in the pterosaur it is the fourth finger that supports the wing; in the bird it is mainly the second, and in the bat it is the second through the fifth. In each animal the wing attaches to the trunk by means of the shoulder girdle, a ring of bones. The girdle of the larger pterodactyls is peculiar in that the scapula, or shoulder blade, turns inward and abuts the notarium, a unique pterosaurian bone, at the midplane of the body. The notarium is several vertebrae fused together. The arrangement provided a base for the action of the wing. The arm of man is shown in *d*.

FIG. 1. Arms and fingers into wings (from Langston[12], courtesy of Scientific American).

2. The arctic tern migrates from the Arctic to the Antarctic, covering thousands of miles between landfalls and staying aloft days and weeks at a time.

3. The ruby-throated hummingbird migrates from Florida to Central America across the Gulf of Mexico.

4. Some swifts stay aloft day and night.

B. High flights

Aircraft can exceed an altitude of 80,000 feet. Birds can't compete. Eagles have been seen higher than 30,000 feet, but birds obtain their sustenance only from locations on or near the ground, and there has been little evolutionary pressure to achieve high-altitude flight.

C. Navigation

Aircraft navigate over long distances using dead reckoning, the magnetic compass, many radio aids, and even inertial navigation.

Birds home over considerable distances, and also navigate effectively during long-distance migrations. They apparently use a variety of clues and senses, including visual geographic landmarks, celestial objects, sky polarization, and magnetic fields. Most amazing, some migrations are conducted without the bird having had experience for the particular flight. The destination must be genetically preprogrammed into the bird's brain; this can be considered the biological equivalent of the navigation of a cruise missile, except that the brain does the job with less material and more versatility.

It has been reported that birds do weather forecasting to pick the right phase of a meteorological pressure system that will provide the tail winds needed to make the migration possible.

Some birds use echo-location over short distances, and of course most bats do. This is the acoustic equivalent of radar.

D. Flight maneuvers

Aircraft are flown in formation, perform aerobatics, and engage in dog fights.

Birds sometimes do the same. The formation flying of ducks and geese saves energy and probably serves some function of social communication as well. Some birds seem to do aerobatics just for the fun of it. I have observed a raven doing fast rolls that had no obvious relation to saving energy or acquiring food. I watched a frigate bird climb to a cloud base in a thermal, far too high to seek food, and then tumble to low altitude like a flopping, limp rag, an inelegant descent mode that looked like "just for the fun of it." In East Africa the Battleur eagle, with a miniscule tail, will intentionally somersault in flight.

As for aerial dog fights, these are common. Hawks will fight each other, and they attack prey in the air as well as swooping down on ground-based prey. Small birds fight off hawks to protect territory.

When birds, bats, and large insects catch small insects in flight, the detections and the maneuvering on both sides resemble aerial combat with aircraft. There are stealth techniques, camouflage, counter-measures, and communications.

E. Aerodynamics; airfoils

As airfoils have evolved by human engineering over the past hundred years, there have been huge improvements in understanding and controlling boundary layers, increasing lift, decreasing drag, and improving pitching moments. Man has refined geometry; added slats; provided flaps and slots and multi-element airfoils; installed boundary-layer trips, vortex generators, and fences; applied spoilers; and utilized variable geometry to adapt a configuration to varying conditions.

Birds have engineered all of the same airfoil features as appropriate for the Reynolds numbers of 10,000 to 200,000 at which their flying surfaces generally operate. At lower Reynolds numbers, the birds' solutions may be better than man's. The frigate-bird observations cited earlier suggested that a lift coefficient $C_L \approx 1.8$ is achieved at a Reynolds number of about 50,000. Pennycuick[13, 14], for a vulture, and Tucker and Parrott[15], for a falcon, suggest that $C_L = 1.6$ is probably obtainable, but the measurements are not definitive. For slope-soaring birds, Pennycuick[16] computed $C_L = 1.63$ for frigate birds and 1.57 for black vultures. Eggleston and Surry[17] made wind-tunnel tests on computer-designed airfoils for model airplanes and found a maximum C_L near 1.6 for Reynolds numbers in the range 34,000-50,000, with the highest C_L being 1.76 at $Re = 36,000$. Carmichael[18] and Pressnell and Bakin[19] show the influence of boundary-layer trips and invigorators on maximum C_L, and obtain experimentally $C_L \approx 1.7$ for an airfoil at a Reynolds number of 30,000. Dilly[20] cites maximum C_L's reaching 1.8 (with very high drag) at Reynolds numbers up to 60,000. Carmichael[18] points out wide variations in measured characteristics in different wind tunnels. Thus the available observations on maximum C_L for both birds and artificial wings at these Reynolds numbers are not definitive, but the data suggest that nature's designs can be as good as the best of man's.

The cross-section of a vulture's primary feather in Fig. 2 shows a very sharp leading edge. Wainfan[21] has investigated how the design of effective airfoils depends on Reynolds number. He concludes that the leading-edge radius should be less than 0.25 percent of the chord at the Reynolds number of 35,000 at which this feather often operates. The observed leading edge fits this criterion. Wainfan notes:

> "Reducing the radius of the leading edge
> turbulates the boundary layer early and
> helps keep the flow attached. This effect is

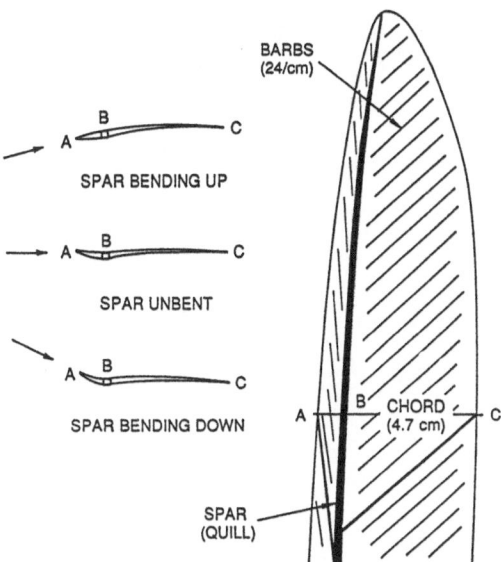

FIG. 2. Vulture primary airfoil; modifications under load. As load bends feather down, *B* moves farther than *A* or *C*. Torsion also rotates feather.

enhanced if the airfoil has relatively little curvature in the first 5% or so of the chord aft of the radius. Turbulating the boundary layer by sharpening the leading edge increases the maximum lift of the airfoil and decreases its drag dramatically."

A striking example of the effectiveness of nature's capability in aerodynamic design is provided by Figure 3, which shows the cross section or airfoil of an insect's wing[22]. It is basically a thin sheet, wrinkled seemingly at random (presumably for both structural and aerodynamic reasons). A conventional airfoil is also shown that encloses the envelope of the wrinkled sheet. Wind tunnel results demonstrate that at a Reynolds number of 900 the insect airfoil works as well as the man-made airfoil; at 450, nature's insect airfoil works *better* than the man-made one. An aerodynamicist developing an airfoil for a Reynolds number of 450 would do well to emulate the insect's shape, at least as a starting point. Without knowing about the insect's solution, the aerodynamicist would certainly not have evolved the sort of complex shape the insect found so useful.

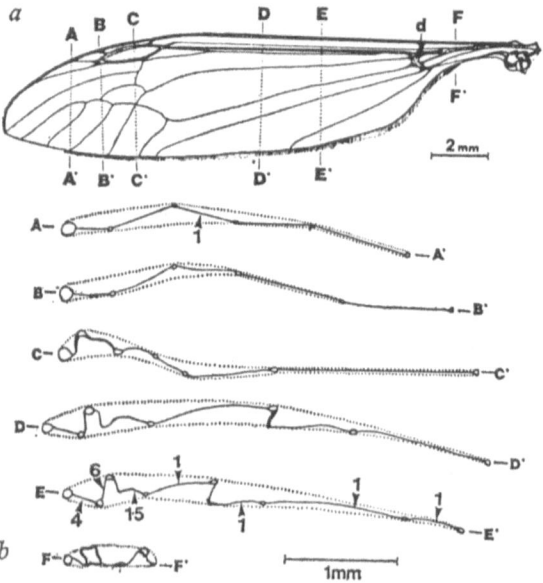

"Fig. 1. a. Left wing of *T. oleracea* (crane fly). d, Cuticular bar which reinforces the corrugation. Section letters apply to *b*. *b*, Chordwise sections of *T. oleracea* wing. Tubular veins to scale, membrane thickness shown diagrammatically. Some measured membrane thicknesses are shown (μm). Dotted lines, section envelopes." .... (from original figure caption[22]).

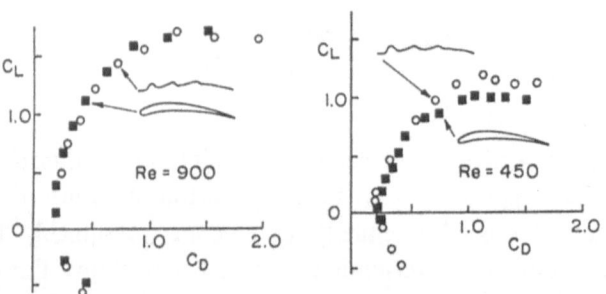

FIG. 3. Architecture and relative aerodynamic performance of insect wings (top portion of figure from Rees[22], courtesy of Nature).*

F. Structure

The variable geometry that is readily obtainable with natural wings is the envy of aerodynamicists and structural engineers. The bat's fingers (Fig. 1) can be controlled for violent maneuvers. The primary feathers that are extended and separated in the wing of a vulture present, all together, a wide-chord, multiple-element airfoil for minimum speed and high maneuverability. Their vertical spread may also diffuse vorticity and benefit induced drag, in a manner somewhat analogous to that of winglets.

Examination of one of the primary feathers of a vulture emphasizes the elegance of nature's design. Figure 2 shows the airfoil of one of the front two primary feathers that bend up in gliding flight and additionally bend up and down from that position in response to down and up flapping. The tapered rectangular-spar cross section is well tailored to fit within the airfoil shape requirements while giving the desired bending strength for up and down loads. The location, centered at about 20 percent chord, provides a twist with lift loads, a load-alleviation technique. The feather construction, with barbs pointing outward toward the tip, acts to vary the airfoil camber with load. Incidentally, an inadvertent separation of the parallel barbs extending out from the spar or shaft is taken care of perfectly by merely touching the previously adjacent barbs together. Tiny hooked filaments lock together and the feather is returned to its original condition; thus Velcro, designed over 100,000,000 years ago.

The hollow-feather spar (quill) construction fits the need for combining strength with lightness. Some bird-wing bones are hollow tubes with lightweight internal truss construction to prevent buckling of the thin walls. The wing bones of large pterodactyls are tubes with extremely thin walls, a construction consistent with low wing loadings in spite of the large wing span. In nature, the wing bones handle all wing torsion, bending, and shear loads; the feathers or membranes provide the aerodynamic shape. With the vertebrates, there is no D-tube construction for handling torsion loads. The exterior would weigh too much. This design is analogous to that of lightweight aircraft; for wing loadings under 10 kg/m^2, the spar (and perhaps associated struts or wires) handles all the main loads, with an airfoil shape as a surrounding "glove" (for example, the Solar Challenger[23]). Higher wing loadings are usually associated with stressed-skin construction to handle torsion.

The variable geometry of birds' wings and tails permits the configuration to be adapted to various needs, both by muscle and by aeroelastic effects. The innate sensing system and brain also tell the muscles how to alter the configuration. Thus the bird can readily adjust to strong turbulence during landing. A rigid-wing airplane cannot; the airflow angles can exceed those that can be handled by slats and flaps and ailerons. There have been various attempts with aircraft to use twist in portions of wings. For example, for the Guggenheim Safe Airplane Competition, in 1929, the Curtiss "Tanager" used floating ailerons to achieve controllability even beyond stall of the fixed biplane wings. The technique worked. A review of the subject and a modern variation are presented by Jones[24]. In 1983 we successfully tested a model glider in which the outer half of each wing was automatically rotated in pitch by an attached tail to maintain a specific lift coefficient. With a roll-rate-controlled servo providing the "brain," the model functioned satisfactorily in turbulence.

Natural wings of vertebrates all fold for maneuvering on the ground, analogous to the folding of wings on carrier aircraft.

A final great structural feature of birds is the retractable landing gear, which is also well adapted to rough-field operation.

G. Instrumentation

In addition to the sensing techniques required for navigation, birds have the ability to monitor flight conditions in ways analogous to airplane instrumentation. They certainly observe attitude and altitude and speed. They are masters at locating thermals and moving to the strongest portion. The methods by which they find and use thermals are not understood, except that sometimes one bird will observe another catching a thermal and fly over to use the same thermal (as sailplanes use other sailplanes as thermal indicators, and occasionally sailplanes use soaring birds, and birds use sailplanes). Birds' "instrumentation" demonstrates remarkable effectiveness. A hawk or vulture may cruise kilometer after kilometer at 20 m above trees, without wing flapping, but with a great deal of maneuvering to translate turbulent ups and downs into sustaining lift. The "instrumentation" is also remarkable in that it permits birds to grab insects and other objects out of the air, to swoop down to the ground or water surface and snatch a mouse of minnow, or, in the case of the pelican, to plunge accurately into the water to catch a fish below the surface. The biological flier also certainly has the ability to monitor "engine" temperature and assess the amount of "fuel" on board. The visual acuity of hawks for finding prey is legendary. Vision may be one element of the sensitive rate-of-climb indicator that a bird must have for assessing upcurrent details or establishing relative height by parallax as the bird moves along its flight path; but this cannot be the explanation for a frigate bird finding lift in a gentle convective cell over a featureless sea.

H. Stability, control, maneuvering

As with all animals, birds have effective active control systems. For an extremely efficient soarer such as the albatross or gannett, during normal flight there is essentially no tail (the body is simply a streamlined low-drag shape) and yet probably the undercambered wing has a negative pitching movement and so is unstable. Small forward and backward movements of the wing may provide an active-control method to produce stable flight. Pennycuick and Webbe[25] report that "moulting fulmars can fly and maneuver at normal and high speeds just as well without a tail" and ascribe pitch control to fore and aft movements of the wings.

During transient maneuvers such as taking off, landing, evading predators, and collecting food on the wing, complex control is manifested, involving large motions of wings and tail, and even body and legs. In a moderate wind, some birds (especially small hawks) are seen to remain in an accurately fixed position relative to the ground. Movies of such flights suggest that the head and eyes can remain rigidly positioned with an accuracy of a millimeter or so, even while the rest of the bird is moved violently about. The head is thus a rigorously stabilized platform. The wings can be controlled to provide yaw forces, and in some birds the horizontal tail, with lift on it, is rotated on a longitudinal axis to augment yaw control (as with the Gossamer Condor).

V. Quetzalcoatlus Northropi (QN)

In late 1983, I felt that advances in composite construction, robotics, stability and control, sensors and servos, and the theory of oscillating airfoils had reached the point where one could make a full-size, radio-controlled, wing-flapping flying replica of QN, the largest natural flier known, with a span of 11 meters. Some flightless birds were much heavier, and the flying teratorn (a six-million-year-old fossil suggests a bird perhaps resembling a condor, with a 7.5-m span) was probably somewhat heavier, but QN had the largest span. The National Air and Space Museum has sponsored our initial studies for re-creating the QN.

The fossil evidence for QN is meager, and so most details of its shape and appearance involve conjecture. The consensus of participants in a 1984 informal workshop on the subject at Caltech was that QN resembled a mixture of an albatross, a frigate bird, and a crane, enlarged more than six-fold. Figure 4 shows the consensus dimensions of QN. The next section considers questions of size vs power.

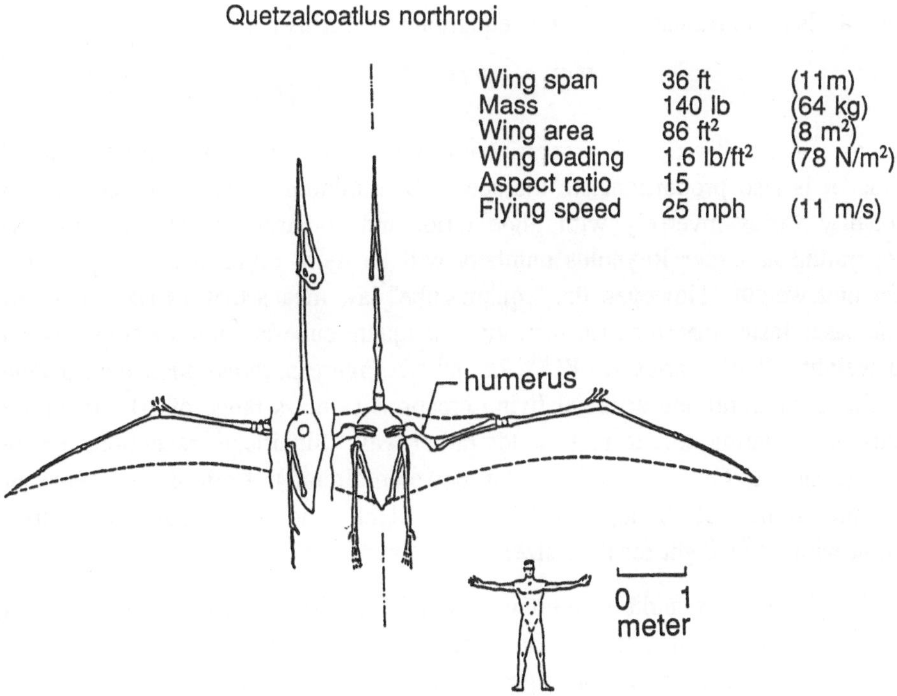

Quetzalcoatlus northropi

Wing span	36 ft	(11m)
Mass	140 lb	(64 kg)
Wing area	86 ft^2	(8 m^2)
Wing loading	1.6 lb/ft^2	(78 N/m^2)
Aspect ratio	15	
Flying speed	25 mph	(11 m/s)

FIG. 4. Plan view and side view of *Quetzalcoatlus Northropi*.

The special problem with QN is to handle the pitch stability/control challenge of a tailless flier, with a cambered-membrane wing that probably had a negative pitching movement. In early 1985 we have been flying a 2.5-m span radio-controlled model with swinging-wing control instead of elevator control. We are edging cautiously toward active control based on use of an angle-of-attack vane and a pitch-rate gyro as sensors to handle pitch stability when the horizontal stabilizer has been completely removed. As pterosaurs evolved to the larger sizes, for which tails disappeared as efficiency improved, the relative size of the brain increased, presumably partly in response to the increased demands of active control.

Eventual flight of the full-size QN will bring back to life the long-extinct flying reptile, and will help to interest people in evolution and in nature's engineering capabilities. The species did not survive the "great extinction" about 63,000,000 years ago, but nevertheless it probably was a success for much, much longer than man.

VI. Scaling laws

If we ignore Reynolds-number effects for a flight vehicle of a specific shape but varying size (defined, say, by span b) and varying weight W, the power to fly, P, is given by

$$P \sim W^{3/2} b^{-1} \ .$$

Since area A is proportional to b^2, this equation transforms to

$$P/W \sim (W/A)^{1/2} \ .$$

Thus the power per unit weight is proportional to the square root of wing loading, and it follows that it is also proportional to the speed for minimum power. The constant of proportionality varies inversely with glide ratio, and so large vehicles (nature's or man's) operating at larger Reynolds numbers will be more efficient and require less power per unit weight. However, the "square-cube" law means that, as size increases, weight increases faster than area (in fact, given a square-cube-law relationship between area and weight, P/W varies as $W^{1/6}$ or $b^{1/2}$). Figure 5 shows that, over a huge range of size for natural and artificial flying creations (a mass range of 10^{11} to 1), the square-cube law relating area to mass holds rather well. Significant exceptions are the pteranodon (related to QN) and HPA (human-powered aircraft). Both of these types are crowding the limits of biologically-powered flight, and so feature specialized construction with low weight for their size.

What the foregoing remarks suggest is that if large birds fly satisfactorily at wing loadings of 5 to 15 kg/m^2, and if man can produce power-to-gross-weight ratios comparable to those of birds, human flight should also be possible at comparable wing loadings. The 30-kg human-powered Gossamer Albatross, with a 65-kg pilot, has a

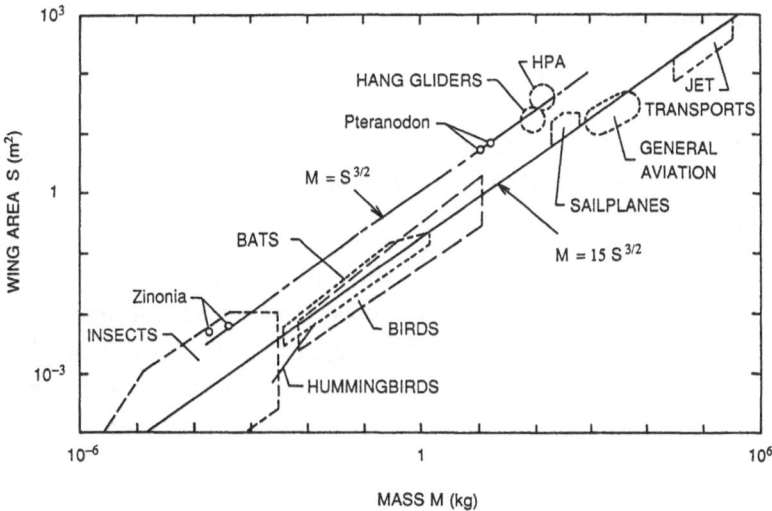

FIG. 5. The square-cube law; variation of wing area with mass (from McMasters[3]).

lower wing loading, about 2 kg/m², than large birds. For birds, flight muscle is about one fourth of total weight, while for a human-powered vehicle the flight muscle is closer to one tenth of total weight. However, for a short time a human can put out two or three times the equilibrium power and so can be comparable to a bird, while the larger human-carrying vehicle can be more efficient due to the aerodynamic and structural options available to the human engineer. Taking all of the above points into consideration, human-powered flight is not surprising. On the other hand, an extrapolation from the flight of soaring birds suggests that the flight of QN is surprising, if the square-cube law holds, because a six- or seven-fold increase in b would require a proportional increase in wing loading and a 2.5-fold increase in P/W, far more than could be made up for by slight aerodynamic improvement at increased Reynolds number. Probably the wing loading of QN was very little higher than that of large soaring birds, of the vulture and eagle variety, and hence a biologically reasonable P/W was maintained.

These deliberations suggest that man is not the only non-flying animal that can maintain flight through muscle power if a suitable vehicle is provided. Animal-powered flight using a dog, a mouse, or even a fish is conceivable, with the size of the flight vehicle being smaller than the size for which the square-cube law inhibits flight. At the small-size end of the animal-powered flight spectrum, the efficiency arising from lower wing loading (and perhaps from a higher P/W for short times, derived from faster metabolism) is fighting the decreased aerodynamic efficiency due to Reynolds-number effects. In summary, the dog-powered or rat-powered airplane is technologically feasible, the main problems being the psychology of the animal and of animal lovers.

VII. Nature vs man as engineer

Attention to biological flight, whether by birds, bats, pterosaurs, or humans, emphasizes low-power aerodynamics and efficient structures. For manned and unmanned aircraft using fossil fuel, the stimulus of biological flight can be to raise the sights of designers. Non-refueled piloted flights with durations in weeks are achievable with existing technology.

Mother Nature has been designing birds and other flying creatures for at least 150,000,000 years. The task has obviously been well done. Every design feature meets a survival purpose, some by way of aerodynamic efficiency, others by way of biological adaptability or sexual selection. Thus, birds have elegantly handled aerodynamic problems such as boundary-layer control, stability, and sensors. Basic study of the aerodynamic features of birds can be expected to yield valuable dividends to both research aerodynamicists and engineers.

References

1. *Symposium on Swimming and Flying in Nature*, Proceedings of the Caltech Symposium of July, 1974, edited by T.Y. Wu (Plenum Press, New York, 1975).

2. A.M. Kuethe, "Prototypes in nature; the carry-over into technology," Technical Engineering Review, University of Michigan, Spring, 3-20, 1975.

3. J.H. McMasters, "Reflections of a paleoaerodynamicist," AIAA 2nd Applied Aerodynamics Conference, Seattle, Paper No. 84-2167, 1984.

4. A.H. Woodcock, "Soaring over the open ocean," Soaring **6**, Nov.-Dec. (1942), reprinted from Scientific Monthly **1942**, Sept.

5. P.B. MacCready, "Soaring bird aerodynamics -- clues for hang gliding," Ground Skimmer **1976**, Oct.

6. P.B. MacCready, "Flight on 0.33 Horsepower: The Gossamer Condor," AIAA 14th Annual Meeting, Washington, D.C., Paper No. 78-308, 1978.

7. P.B. MacCready, "Flight of the Gossamer Condor," Science Year, *World Book*, (Childcraft International, 1979), pp. 85-99.

8. J.D. Burke, "The Gossamer Condor and Albatross: A case study in aircraft design" (AIAA Professional Study Series), AeroVironment, Inc., Report AV-R-80/540, 1980.

9. M. Grosser, *Gossamer Odyssey* (Houghton-Mifflin, Boston, 1981).

10. "Children of Icarus," *Nova, Adventures in Science* (Addison-Wesley, Reading, 1983).

11. P.B. MacCready, "Soaring aerodynamics of frigate birds," Soaring **48**, July, 20-22 (1984).

12. W. Langston, Jr., "Pterosaurs," Scientific American **244**, Feb., 122-136 (1981).

13. C.J. Pennycuick, "Gliding flight of the white-backed vulture *gyps africaur*," J. Exp. Biol. **55**, 13-38 (1971).

14. C.J. Pennycuick, "Mechanisms of flight," *Aviation Biology*, edited by Farner and King, Chap. 1 (Academic Press, 1975).

15. V.A. Tucker and G.C. Parrott, "Aerodynamics of gliding flight in a falcon and other birds," J. Exp. Biol. **52**, 345-367 (1970).

16. C.J. Pennycuick, "Thermal soaring compared in three dissimilar tropical bird species; *fregata magnificens, pellecanus occidentelis and corigype atralus*," J. Exp. Biol. **102**, 307-325 (1983).

17. B. Eggleston and D. Surry, "The development of new A/2 airfoils aided by computer," in *Proc. National Free Flight Symp.*, 127-134 (1980).

18. B.H. Carmichael, "Update on high performance model airfoils," *National Free Flight Symp.*, Rep. 34-41, 1981.

19. M.S. Pressnell and M.S.B. Bakin, "Airfoil turbulator and navigators," in *Proc. National Free Flight Symp.*, 103-109 (1982).

20. M. Dilly, "Special award -- Gyorgy Benedeks airfoils," in *Proc. National Free Flight Symp.*, 82-83 (1981).

21. B. Wainfan, "Design of free flight airfoils for minimum sink," *National Free Flight Symp.*, Rep. 9-13, 1984.

22. C.J.V. Rees, "Form and function in corrugated desert wings," Nature **256**, 201-203 (1975).

23. P.B. MacCready, P.B.S. Lissaman, and W.R. Morgan, "Sun-powered aircraft designs," J. Aircraft **20**, 487-493 (1983).

24. R.T. Jones, "Safety of slow flying aircraft," Sport Aviation **1984**, March.

25. C.J. Pennycuick and D. Webbe, "Observations on the fulmar in Spitsbergen," British Birds **52**, 321-332 (1959).

The Role of Cartoons in Turbulence

G. M. Corcos
Professor of Mechanical Engineering
University of California, Berkeley, California 94720

A cartoon was originally an outline or pattern drawn on cardboard to guide the weaver of a rug or tapestry. In more recent times a cartoon has come to mean a caricature, a representation of something or someone that simplifies and deliberately exaggerates certain features of the subject for purposes which, at least in the mind of the cartoon's author, can be summarized as clarification. The constructions that I will discuss are cartoons in both senses of the word.

A cartoon is a functional representation, in the sense that the nature of cartoons of the same subject changes with their purpose. For example, a public figure is not drawn in the same way when the object is to hold him up as an example of civic virtue as when the object is to convince the viewer that the rascal ought to be thrown out. Similarly, in physics we are familiar with two strikingly different cartoons of visible electromagnetic radiations. When our intent is to study the interactions of light with matter, we take these radiations to be "very small bodies emitted from shiny substances", as Newton put it, or quantized photons, in modern terms. But when we mean to focus on the propagation of light, we are likely to represent it by waves.

I. Cartoons in fluid mechanics

In fluid mechanics, we make use of at least two types of cartoons.

The first type is exemplified by the few known exact solutions of the Navier-Stokes equations. They apply to problems that have been simplified, geometrically or otherwise. Examples of these are the Squire-Landau point-force solution for a jet or the viscous flow toward a rotating (von Karman) or stationary infinite plane. One may object that these are not at all cartoons, since they are exact solutions of a demonstrably comprehensive and faithful mathematical representation of fluid motion. But this view would miss the central role of examples of this kind as crutches for, or even foundations of, our understanding of other flows for which these solutions apply only locally or approximately.

The second type provides a short cut to the solution of the Navier-Stokes equations -- but a short cut in which the essential physical processes are correctly represented, albeit by a simplified mathematical description. This class of cartoons is far more numerous than the first. When we examine examples of these, we note immediately a certain overlap with asymptotic formulations. For instance, it is plausible, though perhaps not historically verifiable, that Prandtl's boundary layer started out as an inspired cartoon; it seems almost certain that the Oseen approximation did. Yet many useful cartoons cannot be derived from the equations of motion by a limiting process. Those that cannot run the risk of being found less respectable than those that can; since they are not systematically derived, their accuracy is occasionally hard to assess. But their lower standing (which is more evident among applied mathematicians than among engineers) may have more to do with their unpredictable origin than with their merits. They are partly products of fancy, and in this respect they are close cousins of non-scientific cartoons. Examples of these are:

Thwaite's method[1,2] for laminar boundary layers; Wedemeyer's solution[3] for the spin-up from rest of the fluid inside a finite circular cylinder; Stewartson's integral method[4] for the semi-infinite impulsively started flat plate; and von Karman's point-vortex model[5] for the wake of a cylinder. Burgers' model equation[6] for the evolution of a one-dimensional shock wave also belongs to this class of fluid-dynamical cartoons. Perhaps because its mathematical structure serves as the primary paradigm, it is esteemed even in the best circles. On the other hand, Turcotte and Oxburgh's boundary-layer treatment[7] of mantle convection provides an interesting mixture of two schools of cartooning, the asymptotic theory and the thumbnail sketch.

It should be evident from both definition and examples that a good cartoon extracts from the physical situation a definite pattern or outline of the flow, physically tractable and dynamically accurate in its essentials.

II. Cartoons in turbulence

When a flow is turbulent, we are torn between two conflicting views of it. The first lends no face, no pattern, to the details of the flow, but only an infinite variety of instantaneous realizations that we despair to characterize. According to this perspective we should reserve reasonable expectations only to averages. No doubt this view itself is an idealization, but one that has proved singularly difficult to exploit. The second view, which has recently been bolstered by a number of observations, is that what appears as infinite variety, shapeless in space or time, may in fact be a blurred image of the superposition of just a few dynamical events, complex but recognizable, stochastic over long enough times, but with a coherent evolution over shorter (though usefully long) intervals. It is this second view, of course, that inspires cartoons of turbulence. These have been used for more than a century by Taylor[8], Prandtl[9], von Karman[5], Synge and Lin[10], Burgers[11], Townsend[12], Corrsin[13], Tennekes[14], Frisch et al.[15], and many others. We owe an intriguing recent cartoon to Lundgren[16]. Philip Saffman[17] has suggested that a cartoon of turbulence should involve more than one type of elementary motion, and he explored in detail the statistical consequences of several such special flows in his Planck Institute lectures.

It is the prospect of enlarged and ambitious scope, such as the scope Saffman envisioned for cartoons of turbulence, that I find exciting. For, while the number, nature and predictability of successive bifurcations suffered by solutions is the object of massive contemporary research, the typical structure of such solutions between bifurcations seems to me to be at least as important. I have previously expressed the (possibly foolish) hope that, at least for some turbulent flows, a cartoon can be made of a short hierarchy of dynamically likely, nested special motions, and that such a construction could be the basis of a statistical treatment of the flow. In other words, one should eventually average over a suitable ensemble or scramble of the solutions provided by this composite cartoon. There are two suggestions here; the first is that the short term dynamics be approximately represented by the nesting of a few special solutions, an extension of non-uniformly valid local solutions of simpler laminar flows. The second is that their long-term effect be derived by averaging over a suitable ensemble of the nested approximations. The guess is that it is far preferable to average over an ensemble of functions that are somewhat more restricted than the true solutions, than to average over completely unrestricted functions and to try to do the selection only after the averages have been taken[18]. The latter procedure seems either insufficiently restrictive (i.e., everything that is known to be true is taken to be true only in the average), or insufficiently motivated (as in closure assumptions).

The first task, and probably the major one, is to find the constituent cartoons and to scaffold them properly into the small nested set, the Grand Cartoon, that describes the evolution of the flow from some initial conditions over a finite but representative time in a dynamically defensible way.

Perhaps to no one's great surprise, the Grand Cartoon has not been completed for any turbulent flow in time for this conference. For the turbulent mixing layer I have suggested a few of its pieces. But even for that one flow, many other pieces are still missing. In particular, I have neither heard of nor found a convincing and explicit dynamical path to what is believed to be the smallest possible scale of motion. While failing to unveil the Grand Cartoon, I propose to describe a few less ambitious cartoons that are meant to illustrate separate phenomena found in the motion of a turbulent shear layer. Perhaps this will encourage others to climb the scaffolding.

III. A cartoon of Kelvin-Helmholtz instability

The cartoon in Fig. 1 of a free shear layer (which may be density-stratified) was proposed by F. S. Sherman and me eight or nine years ago[19]. Figure 2 shows the outline of a typical spanwise vorticity distribution as the two-dimensional instability develops. Before we take up the idealization, there are a few facts to keep in mind.

First, the spanwise vorticity is initially found within a layer of finite and uniform width. The vorticity is of one sign, and its integral over a suitable length, the

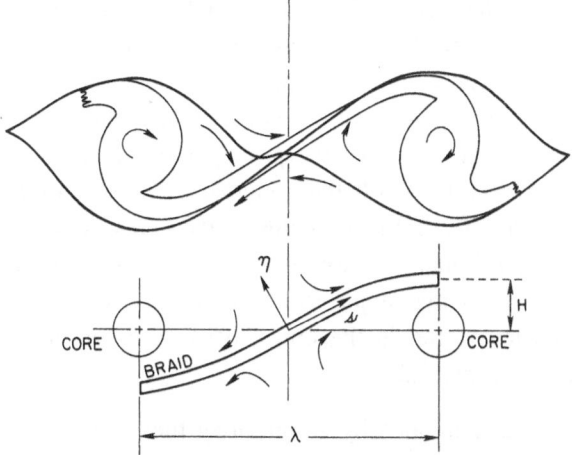

FIG. 1. Kelvin-Helmholtz instability in progress, and its cartoon.

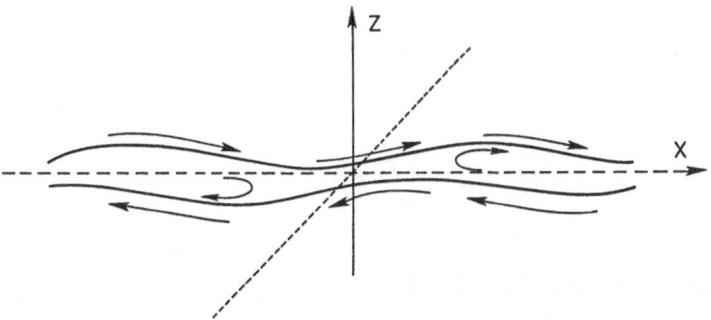

FIG. 2. The initial instability of a shear layer. The accumulation or depletion of vorticity depends on the layer orientation with respect to the positive strain axis.

circulation, remains constant for all times, even if vorticity is generated baroclinically. As a consequence, vorticity can only be redistributed; sources and sinks of vorticity must be of equal strength.

Second, the vorticity distribution is initially only a function of the cross-stream coordinate z. As a consequence, there exists initially an almost perfect, although precarious (in fact, unstable), local equilibrium between strain and vorticity everywhere in the layer. This can be seen by noting that all material surfaces parallel to the center plane of the layer are free from deformation as long as they are undisturbed. Once they are disturbed, the stability problem consists in describing how the vorticity and the associated strain redistribute themselves in space so as to achieve a new *modus vivendi* (it turns out that they never do). A slight periodic excess of vorticity in the regions called the cores in Fig. 1 will cause a slight lateral component of velocity in between the cores. This rotates the central part of the vorticity layer slightly towards the principal axis of positive strain, which makes a 45-degree angle with the streamwise or x-axis.

This process starts a closed chain of events, the Kelvin-Helmholtz instability, which the cartoon will describe. As a result of the rotation, the central part of the vorticity layer experiences a slight positive strain along its length, and this strain chases some vorticity out its ends. Since the total amount of vorticity is conserved, vorticity accumulates in the cores. This accumulation both increases the value of the principal strain rate and further tilts the vorticity-losing layer (the "braid") towards the principal strain axis. Both of these trends accelerate the transfer of vorticity from the braids to the cores. On the other hand, the density of the fluid was originally stratified vertically, but is now stratified across a sloping layer. This leads in the braids to baroclinic generation of vorticity of the same sign as that of the layer, and in the cores, because of vorticity invariance, to generation of vorticity of the opposite sign.

For long waves, if the braids being rapidly thinned by strain along their length are treated as boundary layers (or velocity and density discontinuities), an integral of the vorticity across a braid yields

$$\frac{\partial S}{\partial t} + \bar{U}\frac{\partial S}{\partial s} = -S\frac{\partial \bar{U}}{\partial s} + g^* \sin\theta , \tag{1}$$

where

$$S(s) \equiv u^+ - u^- = -\int_{-\infty}^{\infty} \Omega\, d\eta$$

is the local shear along the braid; $\bar{U} = (u^+ + u^-)/2$; θ is the braid slope angle; and $g^* \equiv g\,\Delta\rho/\rho_0$ is the reduced gravitational acceleration.

The cartoon (Fig. 2) is going to approximate the terms of this equation. It will remain faithful to the word description of the instability. In particular, it will assess approximately how much vorticity is lost by the braids to the cores, how much is regenerated in the braids by baroclinicity, and how much the gain in vorticity increases the amplitude of the waves. But it does so by taking liberties with the representation of the differential advection of vorticity. It assigns all of the vorticity to two separate regions:

a) The cores centered at $x = \pm (1/2 + n)\lambda$, around each of which the circulation is $\Gamma_C(t)$. Here λ is the perturbation wave length.

b) The braids that pass through the points $z = 0$, $x = \pm n\lambda$. These have a circulation

$$\Gamma_B = \int_{-L}^{L} S\, ds$$

and a time-dependent configuration. Vorticity flowing out of the somewhat arbitrary ends $\pm L$ of the braids is transferred to the cores according to the relation

$$\Gamma_B + \Gamma_C = \Gamma = \lambda \, \Delta U = \text{constant.}$$

These exchanges lead, by integration over s, to an ordinary differential equation governing the temporal evolution of Γ_C;

$$\frac{d\Gamma_C}{dt} = \alpha \, \Gamma_B \Gamma_C - 2g \, \frac{\Delta\rho}{\rho_0} \, H \, . \tag{2}$$

The first term on the right represents the advective extrusion rate of vorticity from the ends of the braid into the cores, assuming that the advective speed is proportional to Γ_C while the vorticity present at the end of the braid is proportional to Γ_B. The accuracy of both these postulates has been assessed separately by more ambitious calculation schemes. The second term on the right is a rigorous estimate of the amount of vorticity generated per unit time by baroclinicity in the sloping braid. In this term, the elevation $H(t)$ of the end of the braid above its center can also be taken as a measure of the wave amplitude or vortex radius.

A simple but adequate version of this equation is

$$\frac{dG}{dt} = \omega_1 G - \omega_2 G^2 \, , \tag{2a}$$

where

$$G \equiv \Gamma_C/\Gamma \, ; \qquad \omega_1 \equiv \frac{\pi \, U}{\lambda} \, (1 - \lambda^*) \, ; \qquad \omega_2 \equiv \frac{\pi \, U}{\lambda} \, ; \qquad \lambda^* \equiv \frac{g^* \lambda}{\pi \, \Delta U^2} \, ,$$

and $U \equiv \Delta U /2$ is half the difference between the asymptotic velocities on the two sides of the layer. The solution of Eq. (2a) is

$$G = C \, \frac{\exp(\omega_1 t)}{1 + C \, \dfrac{\omega_1}{\omega_2} \, \exp(\omega_1 t)} \, , \tag{3}$$

where the constant C is fixed by the initial condition

$$C = \frac{4 \, H_0/\lambda}{1 - 4 \, H_0/\lambda} \, .$$

It can be shown that $G/4 \approx H/\lambda$.

What does the cartoon achieve? I think that it does almost all one expects of a theory. It gives approximate answers that are accurately rooted in causes that are easy to comprehend and interpret. Note that Eq. (2a) superficially resembles a Landau equation, but that, unlike the latter, its basis is squarely non-linear and requires no amplitude restriction. Nevertheless, it gives a cut-off wave length for instability

($\lambda^* = 1$) which agrees (presumably accidentally) with linear theory for a long wave. The model solution also selects the wave that can reach the largest amplitude for specified velocity and density jumps across the layer. In addition, an examination of the basis of the model allows the interpretation that the second prediction should also apply in the case of a two-dimensional pairing instability, if this takes place, although in truth the model does not provide a mechanism for the initiation of this instability.

IV. A cartoon for the transformation of vortex sheets into vortex tubes

My second cartoon explores another aspect of the interaction of strain with vorticity. In a turbulent mixing layer, even though the initial flow is two-dimensional, with vorticity perpendicular to the plane of the flow, there are several mechanisms that allow components of vorticity perpendicular to the original direction to appear. All of these mechanisms may not yet be known. At any rate, I do not know of a good cartoon to account for them. But there is a cartoon to show that if the vorticity points in the direction of the flow, and if it is strained in that direction, it will tend to behave as if the strain were axially symmetric, and so will end up as Burgers vortices. In this cartoon (Fig. 3), the deformation is assumed to be planar. The principal axis of positive strain coincides with the constant direction of the vorticity and the strain is uniform. We shall return to these restrictions later.

Figure 4, in which the initial vorticity distribution in (4a) may be imagined as the result of a balance between strain and viscous diffusion, also reveals in (4b) the cause of the evolution. Self-induction rotates the vorticity distribution, while inward strain limits this rotation locally to a finite angle, leaving an unbalanced inward velocity component. This focuses or shrinks the span of the sheet and, because the associated stretching intensifies the local vorticity, allows it to rotate some more (Fig. 5). When the sheet has been sufficiently foreshortened, the vorticity-laden fluid generally spins so fast that its inertia prevents it from responding well to the rapid angular variation of the inward strain which it experiences. Instead, it behaves as if the strain were approximately axisymmetric and had an inward component equal to the average it sees over a revolution.

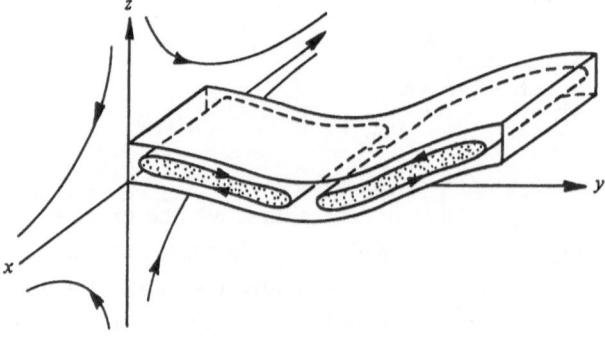

FIG. 3. The cartoon of streamwise vorticity subjected to plane streamwise strain.

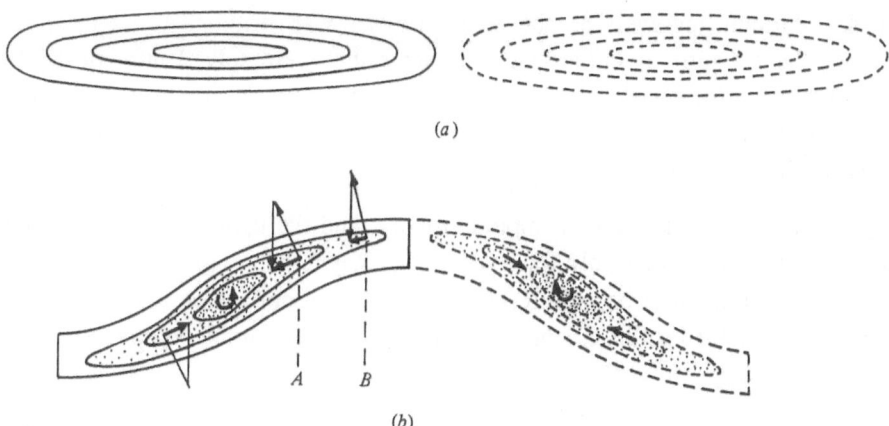

(a)

(b)

FIG. 4. a) A simple strain-diffusion balance. b) Strain, diffusion, and self-induction; early stage.

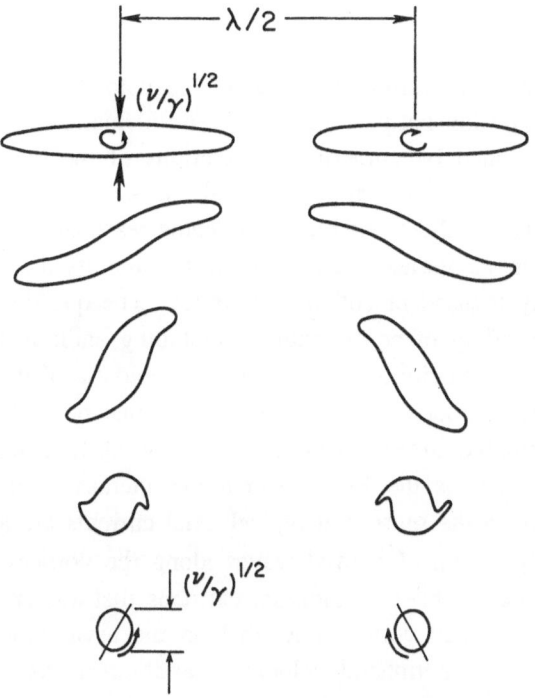

FIG. 5. Collapse of a row of alternating vortex ribbons. The collapse time τ_C is proportional to $\lambda^2 \gamma / \sigma^2$ (γ = strain rate, σ = sheet initial strength).

If the axial strain rate is γ, this is $\gamma/2$, so that the vorticity should acquire approximately the structure of a Burgers vortex.

This word picture, summarized by Fig. 5, can either be translated into a thumb-nail-sketch type of mathematical cartoon[20] or it can be given an asymptotic treatment[21].

Note that this cartoon (as well as the first cartoon) was originally suggested by examination of the outcome of initial-value problems that were solved numerically by the method of finite differences. Note also that the evolution described above would also apply *a fortiori* to a vortex subjected to a strain having two unequal components perpendicular to the vortex axis, both negative.

The numerical experiments, the asymptotic theory, and the cartoon are in thorough agreement. They all tell us that only for a maximum Reynolds number of the order of unity or less can a vortex sheet be spared the eventual fate of collapsing into a tube, and that the time τ_C required for this essentially non-linear collapse to be completed is

$$\tau_C \propto \frac{\lambda^2 \gamma}{\sigma^2}, \tag{4}$$

where γ is the strain rate, σ is the initial strength of the sheet, λ is the length scale for the gradient of σ, and the constant of proportionality is substantially smaller than unity.

It is worth asking whether a spatially non-uniform strain would lead to different results. This is a particularly relevant question because very large strain fluctuations, several times the average strain, are revealed by numerical calculations of the two-dimensional strain that develops in the mixing layer along material surfaces that are likely carriers of streamwise vorticity. Under these conditions, it seems reasonable to suspect that additional phenomena, such as vortex breakdown, might modify in a fundamental way the structure of freshly focused or collapsed vortices. The question can be investigated with the help of a variety of considerations, including additional cartoons[22]. One guiding idea is that if the velocity and strain are characteristic of the large scales of turbulence, the Reynolds number associated with them must be high. Additionally, if one assumes that the circulation around vortices is only a weak function of the Reynolds number, one deduces that, as the Reynolds number increases, the rotational flow is increasingly subcritical, in the sense that typical axial currents are a decreasing function of the speed of propagation of inertial waves along the vortices. Thus vortex breakdown should become impossible. In addition, cartoons that analyze simple idealizations of a non-uniformly strained vortex show that, in the asymptotic limit considered (small axial velocity next to peripheral velocity), variations in axial strain on the appropriate time scale cause no new space scales and only small changes in the structure of the vortex.

New features of the flow may arise from many other quarters, such as bending of vortices, which leads to loops and hairpins that might be represented by tractable analytical models. But let us consider another cartoon in some detail. It is one that describes the interaction between two types of vorticity as an initial value problem. It comes about in this way:

V. A cartoon of vorticity interaction; the spanwise-vorticity amplifier

My second cartoon described the creation of concentrated streamwise vortices with finite circulation and considerable vorticity. Typically, in a mixing layer these would be found in the braids, where spanwise vorticity is almost absent, but also within the cores, where spanwise vorticity is present although it may be a good deal more diffuse than the strain-enhanced streamwise vorticity that is advected out of the braids and wrapped around and in the cores (see Fig. 6). Thus the second cartoon may yield the essential description of the braid. But in the cores we need to inquire how the two types of vorticity interact. A third and last cartoon focuses on a central part of that question -- the effect of a strong streamwise vortex on the spanwise vorticity and streamwise velocity associated with it. The initial state is described in Fig. 7, using obvious geometrical and physical simplifications. The subsequent evolution of this state has been given a systematic and illuminating asymptotic representation by Neu[23]. Another recent solution with a different focus and range of conditions has been published by Pearson and Abernathy[24].

The initial conditions shown consist of a uniform shear,

$$u_1 = \hat{e}_x \Omega z \ ,$$

superposed on an axially-symmetric unstrained vortex,

$$u_2 = \hat{e}_\theta V_\theta \ .$$

The cylindrical coordinates are x, r, and θ, the latter measured from the y axis. Subsequently, $V_\theta (r, \theta, t) = u \cdot \hat{e}_\theta = V_\theta (r, \theta, 0)$, but $U = u \cdot \hat{e}_x$ satisfies

$$\frac{\partial U}{\partial t} + \frac{V_\theta}{r} \frac{\partial U}{\partial \theta} = \nu \nabla^2 U \ , \tag{5}$$

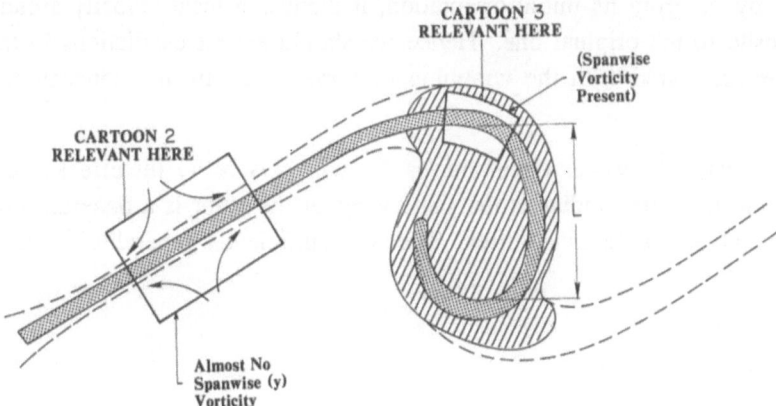

FIG. 6. A sketch of a planar cut through a mixing layer. In the braids, only streamwise vorticity survives; in the cores, streamwise and spanwise vorticity interact.

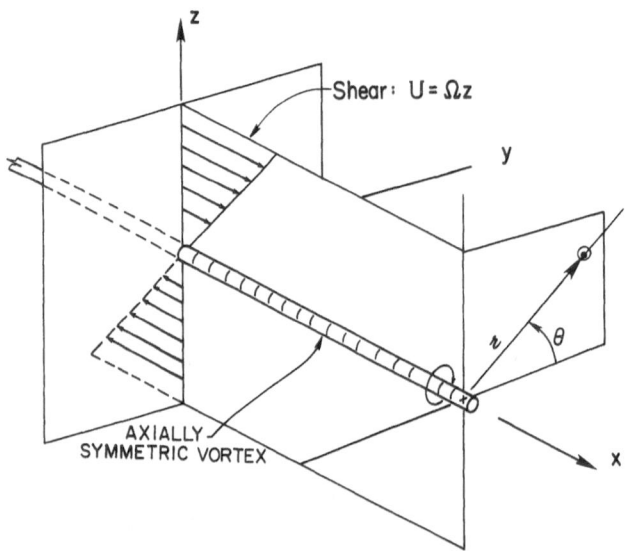

FIG. 7. The simplest cartoon of interaction between streamwise and spanwise vorticity. The initial state of the streamwise velocity profile is shown.

where

$$\nabla^2 = \frac{\partial^2}{\partial r^2} + \frac{1}{r}\frac{\partial}{\partial r} + \frac{1}{r^2}\frac{\partial^2}{\partial\theta^2} .$$

Thus V_θ, the velocity induced by the streamwise vortex, is unaffected by the shear field U. But U is coupled to V_θ in a way which is best visualized by imagining the inviscid evolution of the initially straight vortex lines (parallel to the y axis) that are associated with the initial shear. As shown in Fig. 8, these material lines are eventually wrapped around the streamwise vortex. Now when a vortex line is bent so that its orientation differs by π from its initial orientation, it induces a local velocity around itself exactly opposite to the original one. Hence we should expect oscillations in the shear $U(r)$ to be associated with the wrapping of vortex lines by the concentrated streamwise vortex.

It can be shown that if the vortex circulation is Γ, with $\Gamma/\nu \gg 1$, the effect of the vortex on U within the vortex radius is the same whether the latter is a potential line vortex or a diffusive Oseen or Rankine vortex. Thus it is sufficiently general to assume

$$V_\theta = \Gamma/2\pi r$$

for all times. Equation (5) becomes

$$\frac{\partial U}{\partial t} + \frac{\Gamma}{2\pi r^2}\frac{\partial U}{\partial\theta} = \nu\,\nabla^2 U , \qquad (6)$$

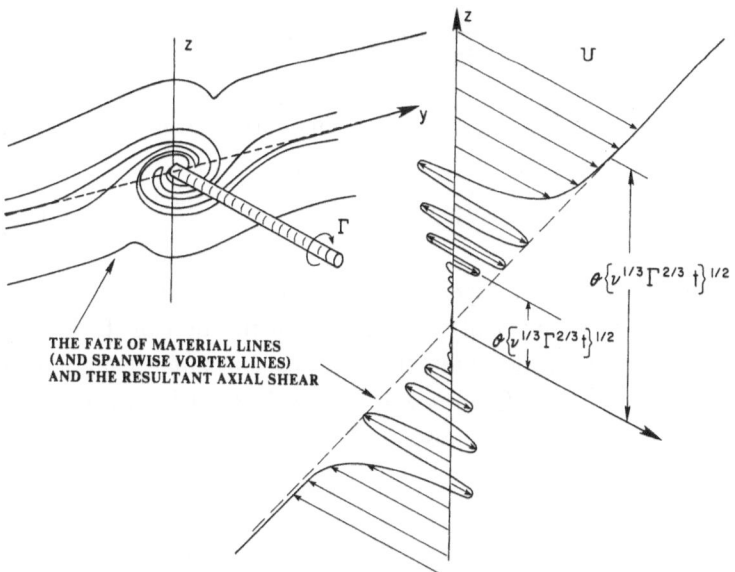

$\theta \{\nu^{1/3} \Gamma^{2/3} t\}^{1/2}$

$\theta \{\nu^{1/3} \Gamma^{2/3} t\}^{1/2}$

THE FATE OF MATERIAL LINES
(AND SPANWISE VORTEX LINES)
AND THE RESULTANT AXIAL SHEAR

FIG. 8. The amplified and fluctuating shear is caused by the streamwise vortex, which wraps spanwise vortex lines around itself.

with boundary and initial conditions

$$U \to \Omega z \quad \text{as} \quad r \to \infty; \qquad U = \Omega z \quad \text{at} \quad t = 0 . \qquad (7)$$

The inviscid version of Eq. (6) can easily be solved by the method of characteristics;

$$U = \Omega r \, \sin \left(\theta - \frac{\Gamma t}{2\pi \, r^2} \right) . \qquad (8)$$

The characteristics are given by

$$\theta = \theta_0 + \frac{\Gamma t}{2\pi \, r^2} , \qquad (9)$$

where θ_0 is a constant, either the value of θ at the initial instant or the asymptotic value as $r \to \infty$. When the flow is inviscid, the value of U oscillates smoothly, along any fixed radius θ, between the extrema $\pm 2\Omega r \sin \theta_1$, which are reached when

$$\left(\theta_1 - \frac{\Gamma t}{2\pi \, r^2} \right) = (1/2 \pm n) \, \pi \, ; \qquad n = 0, 1, 2, 3, \cdots \qquad (10)$$

whereas $U = 0$ on the two spirals

$$\theta = (\alpha + \frac{\Gamma t}{4\pi r^2})$$

where $\alpha = 0$ or π. The radial spacing between two neighboring arms of the same spiral is

$$\Delta r = O\{r^3/\Gamma t\} . \tag{11}$$

Note that the solution of Eqs. (6) and (7) depends on θ and on a similarity variable $s = r/(\delta t)^{1/2}$, where $\delta = v\,h(\Gamma/v)$ and $h(\beta)$ is any smooth function of β chosen for scaling purposes. The estimate (11) shows that the spiral becomes increasingly tight as r decreases. Since we wish s to be $O\{1\}$ where the spiral spacing is about equal to a typical diffusion thickness $(v\,t)^{1/2}$, we choose

$$\delta \equiv \Gamma^{2/3}\,v^{1/3} , \tag{12}$$

and note that, where $s = O\{1\}$,

$$r = O\{\delta t\}^{1/2} = O\{\Gamma^{1/3}\,v^{1/6}\,t^{1/2}\} . \tag{13}$$

At this radius, the inviscid estimate for U is

$$U = O\{\Omega\,(\delta t)^{1/2}\} .$$

We thus define

$$U \equiv \Omega\,(\delta t)^{1/2}\,\hat{u}(s,\theta); \qquad \delta \equiv \Gamma^{2/3}\,v^{1/3}; \qquad s \equiv r/(\delta t)^{1/2}; \qquad \varepsilon \equiv (v/\Gamma)^{1/3} .$$

Then Eq. (6) becomes

$$\frac{1}{2}\left[\hat{u} - s\,\frac{\partial\hat{u}}{\partial s}\right] + \frac{1}{2\pi\,\varepsilon\,s^2}\,\frac{\partial\hat{u}}{\partial\theta} = \varepsilon^2\left[\frac{\partial^2\hat{u}}{\partial s^2} + \frac{1}{s}\,\frac{\partial\hat{u}}{\partial s} + \frac{1}{s^2}\,\frac{\partial^2\hat{u}}{\partial\theta^2}\right] . \tag{14}$$

Now, according to Eq. (8), the inviscid variation of $U/\Omega r$ is nil along the characteristics. In the asymptotic case, $\Gamma/v \gg 1$, we should be able to approximate the elliptic problem defined by Eqs. (6) and (7) by a parabolic one in a coordinate system for which one of the coordinates is constant along the characteristics. Consequently we choose as new (non-orthogonal) coordinates s and θ_0, where, according to Eq. (9), θ_0 is the asymptotic value of θ along a given characteristic and thus designates the characteristic. In terms of s and θ_0, Eq. (14) becomes

$$\frac{1}{2}\left[\hat{u} - s\,\frac{\partial\hat{u}}{\partial s}\right] = \varepsilon\,\nabla^2\hat{u} , \tag{15}$$

where

$$\nabla^2 = \left[\frac{\partial}{\partial s} + \frac{1}{\pi \varepsilon s^3} \frac{\partial}{\partial \theta_o} \right]^2 + \frac{1}{s} \left[\frac{\partial}{\partial s} + \frac{1}{\pi \varepsilon s^3} \frac{\partial}{\partial \theta_o} \right] + \frac{1}{s^2} \frac{\partial^2}{\partial \theta_o^2} \tag{16}$$

is the Laplacian in (s, θ_o) coordinates. Since the inviscid solution is continuous and smooth for $s \gg 1$, the diffusion thickness is

$$\Delta d = O\{v\,t\}^{1/2} = O\{\Delta r / s^3\} .$$

As $s \to \infty$, viscous stresses become negligible, so that the solution must tend to that given by Eq. (8). This limit yields the boundary condition

$$\hat{u} \sim s \sin \theta_o \quad \text{as} \quad s \to \infty . \tag{17}$$

This boundary condition obviates the need for an asymptotic expansion for the outer part of the spiral. Alternatively, use of an outer expansion (for the loosely wound part of the viscous spiral) and matching with the inner expansion reveals that the inner expansion $(s = O\{1\})$ is uniformly valid.

In the limit $\varepsilon \to 0$, the leading-order approximation to Eq. (15) is the parabolic equation

$$\frac{1}{2} \left[\hat{u} - s \frac{\partial \hat{u}}{\partial s} \right] = \frac{1}{\pi^2 s^6} \frac{\partial^2 \hat{u}}{\partial \theta_o^2} . \tag{18}$$

The solution of Eqs. (17) and (18), by separation of variables, is

$$\hat{u} = s \, \exp(-1/3\pi^2 s^6) \sin \theta_o ,$$

or

$$U = \Omega \, r \, \exp[-(\delta t)^3/3\pi^2 r^6] \sin \left[\theta - \frac{\Gamma t}{2\pi r^2} \right] . \tag{19}$$

For small values of s, the exponential term gives negligible values for U, because the positive and negative values in the inviscid approximation alternate over such small radial increments at a given θ that viscosity erases them effectively. Farther out ($s = O\{1\}$), the excursions of U are attenuated but not erased. Still farther out ($s \gg 1$), the inviscid solution is recovered.

The variation of U with r for $\theta = \pm \pi/2$ is sketched in Fig. 8.

VI. Cross-stream vorticity

Let the vorticity be $\omega = \hat{e}_x \omega + \omega'$, where ω is the component associated with the prescribed streamwise vortex and

$$\omega' = \hat{e}_y \frac{\partial u}{\partial z} - \hat{e}_z \frac{\partial u}{\partial y} .$$

We refer to ω' as the cross-stream vorticity and note that it can readily be obtained from Eq. (19). Thus, as $\varepsilon \to 0$, and to $O\{\varepsilon^{-1}\}$,

$$|\omega'| \sim \frac{\Omega}{\pi \varepsilon s^2} \exp(- 1/3\pi^2 s^6)| \cos \theta - \frac{1}{\pi \varepsilon s^2}| . \tag{20}$$

For $\varepsilon = 1/30$, this quantity is shown as a function of s along a radius $\theta = \pi/2$ in Fig. 9. Note that the maximum amplification of the spanwise vorticity by the streamwise vortex is about $(2\varepsilon)^{-1}$.

Thus large spikes of cross-stream vorticity have been created by the streamwise vortex, and these are associated with a rapidly oscillating shear profile. The radial

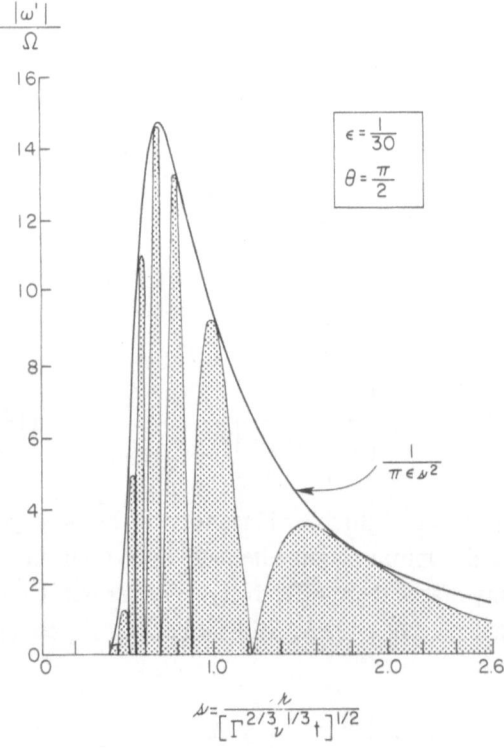

FIG. 9. The magnitude of the component of vorticity perpendicular to x as a function of non-dimensional radius along an arbitrary radial direction.

region over which this large amplification of vorticity has occurred is about $(\Gamma^{2/3} \nu^{1/3} t)^{1/2}$; i.e., larger by a factor $O\{1/\varepsilon\}$ than the radius that the primary vortex would have acquired if it had been allowed to diffuse. So the perturbations are large, widely spread, and have a relatively small spatial scale. We may express our results in terms of the parameters of a mixing layer in Fig. 10, where we have taken as L the diameter of a typical spanwise vortex and we have assumed that the circulation around a streamwise vortex is

$$\Gamma = \alpha \, U L \ ,$$

α being a constant of order unity. As Fig. 11 suggests, the newly created vorticity layers are surely highly unstable and will create ring-like vortices in a very short time; i.e., $O\{|\omega'|^{-1}\}$. These in turn must have an interesting (although so far unexplored) evolution.

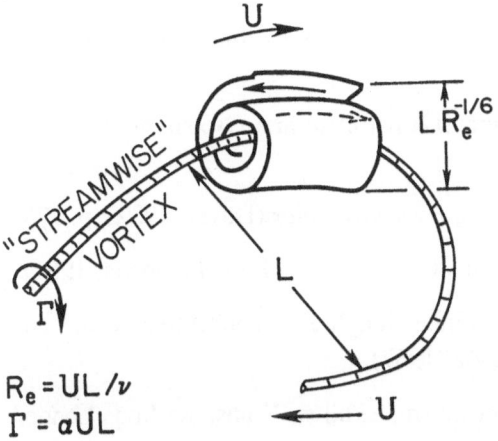

$$R_e = UL/\nu$$
$$\Gamma = \alpha UL$$

FIG. 10. Amplification of cross-stream vorticity. The new shear layers have thickness $d \sim L Re^{-1/2}$; shear $U' \sim U Re^{-1/6}$; vorticity $|\omega'| \sim (U/L) Re^{1/3} \sim 1/\text{growth time for}$ instability; and local Reynolds number $\sim Re^{1/3}$. If new streamwise vortices are created, their diameter $\sim (\nu/\gamma)^{1/2} \sim L Re^{-2/3}$.

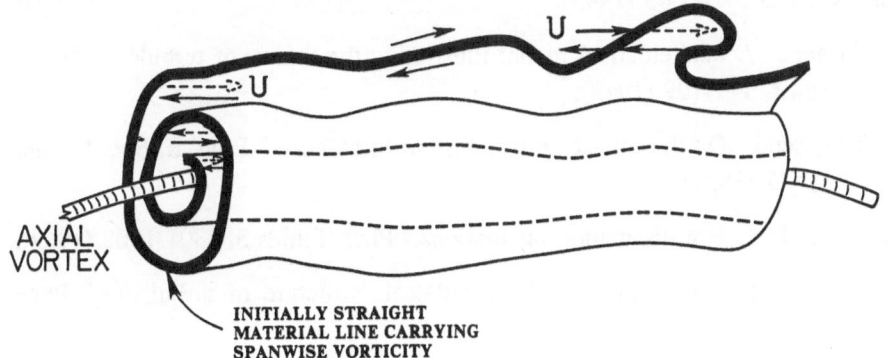

FIG. 11. A sketch of the layers of alternating streamwise velocity, suggesting their susceptibility to shear instability.

In conclusion, cartooning in turbulence is not at all new. But experiments are providing us with a fresh focus and giving us powerful encouragement to continue our cartooning. Meanwhile, numerical experiments are supplying the cartoonist with both qualitative and quantitative guidance. But at the same time they are holding the cartoonist to much stricter standards of dynamical plausibility.

References

1. B. Thwaites, "Approximate calculations of the laminar boundary layer," Aeron. Quart. **1**, 245-280 (1949).

2. B. Thwaites, "On the momentum equation in laminar boundary layer flow under adverse pressure gradients," Aeron. Res. Council, R&M 2514 (1952).

3. E.H. Wedemeyer, "The unsteady flow within a spinning cylinder," J. Fluid Mech. **20**, 383-399 (1964).

4. K. Stewartson, "On the impulsive motion of a flat plate in a viscous fluid," Quart. J. Mech. **4**, 182-198 (1951).

5. T. von Karman (1911); see, e.g., H. Lamb, *Hydrodynamics* (Dover, 1932), p. 225.

6. J.M. Burgers, *Lecture Notes* (California Institute of Technology, Pasadena, 1951).

7. D.L. Turcotte and E.R. Oxburgh, "Finite amplitude convective cells and continental drift," J. Fluid Mech. **28**, 29-42 (1967).

8. G.I. Taylor, "Eddy motion in the atmosphere," Philos. Trans. R. Soc. London **215A**, 1-26 (1915).

9. L. Prandtl, "Bericht über Untersuchungen zur ausgebildete Turbulenz," Z. angew. Math. Mech. **5**, 136-139 (1925).

10. J.L. Synge and C.C. Lin, "On a statistical model of isotropic turbulence," Trans. R. Soc. Canada **37**, 45-63 (1943).

11. J.M. Burgers, "A mathematical model illustrating the theory of turbulence," Adv. Appl. Mech. **1**, 171-199 (1948).

12. A.A. Townsend, "On the fine-scale structure of turbulence," Proc. R. Soc. London **208A**, 534-542 (1951).

13. S. Corrsin, "Turbulent dissipation fluctuations," Phys. Fluids **5**, 1301-1302 (1962).

14. H. Tennekes, "Simple model for the small-scale structure of turbulence," Phys. Fluids **11**, 669-671 (1968).

15. U. Frisch, P.-L. Sulem, and M. Nelkin, "A simple dynamical model of intermittent fully developed turbulence," J. Fluid Mech. **87**, 719-736 (1978).

16. T.S. Lundgren, "Strained spiral vortex model for turbulent fine structure," Phys. Fluids **25**, 2193-2203 (1982).

17. P. Saffman, "Lectures on homogeneous turbulence," in *Topics in Nonlinear Physics*, edited by N.J. Zabusky (Springer-Verlag, New York, 1968), pp. 485-614.

18. The operation of Reynolds averaging is normally applied after a flow has been deemed "turbulent". The adjective thus turns from descriptive to prescriptive, and any opportunity to study the dynamics of the complex flow disappears along with the time derivative.

19. G.M. Corcos and F.S. Sherman, "Vorticity concentrations and the dynamics of unstable free shear layers," J. Fluid Mech. **73**, 241-264 (1976).

20. S.J. Lin and G.M. Corcos, "The mixing layer: deterministic models of a turbulent flow. Part 3," J. Fluid Mech. **141**, 139-178 (1984).

21. J. Neu, "The dynamics of stretched vortices," J. Fluid Mech. **143**, 253-276 (1984).

22. H.H.C. King, "The effect of non-uniform strain on the diffusion of a scalar and on vorticity," Ph.D. Dissertation, Mechanical Engineering (University of California, 1985).

23. J. Neu, "Streamwise vortices in an ambience of spanwise vorticity" (unpublished).

24. C.F. Pearson and F.H. Abernathy, "Evolution of the flow field associated with a streamwise diffusing vortex," J. Fluid Mech. **146**, 271-283 (1984).

Early Transonic Ideas in the Light of Later Developments

J.D. Cole
Professor of Applied Mathematics
Rensselaer Polytechnic Institute, Troy, New York 12180

I. Introduction

In this paper I want to outline some early ideas about transonic flow with which Hans W. Liepmann was associated, and show how they look in the light of later developments. The earliest ideas can be traced back to studies in gasdynamics; for example, Chaplygin's paper[1] in 1902, which treated planar gas jets by the hodograph method. Modern theoretical work connected to aeronautics dates from the papers of von Karman[2], Guderley[3], and Frankl, who all derived the approximate equation of transonic flow around 1946. Early experimental work was carried out by Stack and Dryden at NASA Langley in the early 1940's.

Karman's paper represented the velocity potential Φ for flow past an airfoil (as in Fig. 1) as a uniform flow at the critical speed a^* plus a small disturbance;

$$\Phi = a^*\bar{x} + \phi(\bar{x}, \bar{y}) \ .$$

Assuming that $\partial/\partial\bar{x} \gg \partial/\partial\bar{y}$, in view of the transonic nature of the flow, Karman derived the equation

$$(\gamma + 1)\frac{\phi_{\bar{x}}}{a^*}\phi_{\bar{x}\bar{x}} = \phi_{\bar{y}\bar{y}} \ . \tag{1}$$

This basic nonlinear equation of changing type is at the heart of all transonic theory. Karman also (effectively) noted the similarity parameter

$$K = \frac{1 - M_\infty^2}{\delta^{2/3}} \ , \tag{2}$$

where

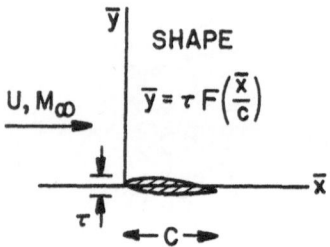

FIG. 1. Transonic flow past an airfoil.

$$\delta = \tau/c \ , \qquad M_\infty = \frac{U}{a_\infty} = \text{Mach number at infinity} = \frac{\text{flight speed}}{\text{sound speed}} \ .$$

Karman provided scaling rules for airfoil flows. Liepmann, Ashkenas, and Cole[4] gave a more detailed derivation and showed that including dilatational viscosity can yield a viscous transonic equation having smooth solutions even when shocks are present. The shocks were shown to be thin; i.e., Re_s = Reynolds number of the shock ≈ 1.

Topics of considerable interest were the possibility of obtaining shock-free mixed subsonic-supersonic flows by use of the hodograph method; the physical significance of the limiting line; and the effects of the viscous boundary layer on the inviscid flow, especially in shock-wave/boundary-layer interactions. These topics are discussed below.

Another significant achievement of von Karman was to bring Hans W. Liepmann to Caltech, where he stimulated and influenced a whole generation of students and created an early interest in transonic flow.

II. Transonic small-disturbance theory

The equations of transonic small-disturbance (TSD) theory can be regarded as part of a systematic limit-process expansion. The starting point is the Euler equation for inviscid compressible flow together with the Rankine-Hugoniot jump conditions for shock waves, including the condition that the entropy can only increase. A typical geometry is shown (in transonic coordinates) in Fig. 2, where a vortex sheet trails downstream from a lifting wing. The asymptotic expansion has the form[5]

$$\frac{q}{U} = (1 + \delta^{2/3} u(x, \tilde{y}, \tilde{z}; K) + \delta^{4/3} u_2 + \cdots) i_x + \delta v + \delta^{5/3} v_2 + \cdots \quad , \quad (3)$$

$$v = (v, w) \quad ,$$

$$\frac{p}{p_\infty} = 1 + \delta^{2/3} P + \cdots \quad ,$$

$$\frac{\rho}{\rho_\infty} = 1 + \delta^{2/3} \sigma + \cdots \quad .$$

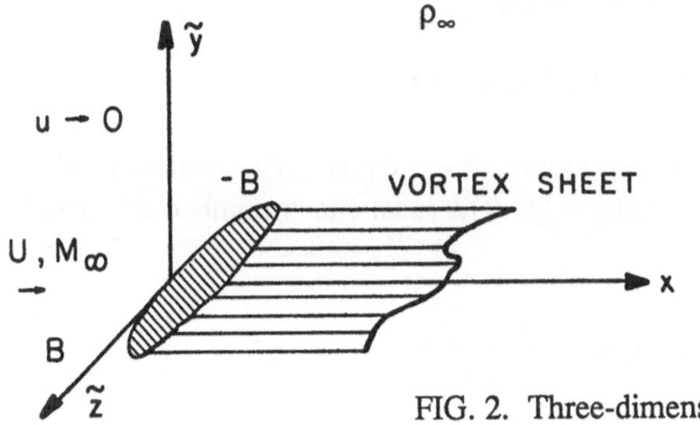

FIG. 2. Three-dimensional wing.

The limit process has $\delta \to 0$, $M_\infty^2 = 1 - K\delta^{2/3} \to 1$, with x, $\tilde{y} = \delta^{1/3}y$, $\tilde{z} = \delta^{1/3}z$, and $K = (1 - M_\infty^2)/\delta^{2/3}$ all fixed. Lengths are measured in terms of a typical wing chord c. The span b should grow as $\delta \to 0$ such that $B = b\delta^{1/3}$ is fixed. Timman[6] and Krupp and Cole[7] showed how these ideas can be extended to unsteady flow, using a dimensionless time coordinate

$$\tilde{t} = \frac{U}{c} \delta^{2/3} t .$$
(4)

The representative point (x, y, z) runs far from the body as $\delta \to 0$, $M_\infty \to 1$ for fixed $(x, \tilde{y}, \tilde{z})$, expressing the fact of large lateral extent for disturbances when $M_\infty \approx 1$. When the limit-process expansion is substituted into the basic Euler system it is found that, to this order, a disturbance potential $\phi(x, \tilde{y}, \tilde{z}, \tilde{t})$ exists such that

$$u = \phi_x, \qquad \boldsymbol{v} = \tilde{\nabla}\phi, \qquad \tilde{\nabla} = \left[\frac{\partial}{\partial\tilde{y}}, \frac{\partial}{\partial\tilde{z}} \right] .$$
(5)

This disturbance potential vanishes at upstream infinity and satisfies the basic TSD equation

$$\left[K - (\gamma + 1)\phi_x \right] \phi_{xx} + \tilde{\nabla}^2\phi - 2\phi_{x\tilde{t}} = 0 .$$
(6)

The pressure coefficient $c_p = 2(p - p_\infty)/\rho_\infty U^2$ is found from

$$c_p = -2\,\delta^{2/3}\,\phi_x .$$
(7)

Some properties of the TSD equation

$$\left[K - (\gamma + 1)\,\phi_x \right] \phi_{xx} + \phi_{\tilde{y}\tilde{y}} = 0 \qquad \text{(TSD equation)}$$
(8)

can now be summarized for steady flow in two dimensions. The equation (8) is of changing type;

$$\text{elliptic if} \quad (\gamma + 1)\,\phi_x < K ,$$

$$\text{hyperbolic if} \quad (\gamma + 1)\,\phi_x > K ,$$

with flow that is locally subsonic or supersonic, respectively. The condition $K = 0$ corresponds to sonic flow at infinity $(M_\infty = 1)$. The local Mach number M_l can be shown to be given by

$$K - (\gamma + 1)\,\phi_x = \frac{1 - M_l^2}{\delta^{2/3}} .$$

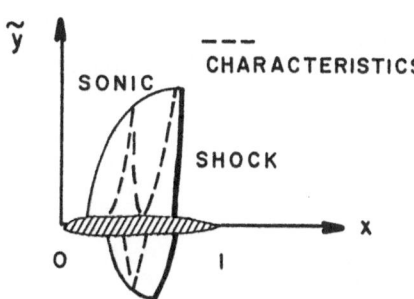

FIG. 3. Flow field for transonic airfoil.

The typical structure of flow at high subsonic Mach number is shown in Fig. 3, where local supersonic (hyperbolic) regions appear over an airfoil. Each supersonic region contains Mach lines or characteristics given by

$$\frac{d\tilde{y}}{dx} = \frac{\pm 1}{((\gamma + 1)\, \phi_x - K)^{1/2}} \tag{9}$$

and is terminated by a shock. The shock conditions are contained in the conservation form corresponding to (8),

$$\left[K\phi_x - \frac{\gamma + 1}{2}\, \phi_x^2\right]_x + (\phi_{\tilde{y}})_{\tilde{y}} = 0 \;, \tag{10}$$

which is a version of the continuity equation. The shock is a discontinuity surface across which ϕ_x and $\phi_{\tilde{y}}$ jump. The jump conditions are given by the integrated form of (10),

$$[K\phi_x - \frac{\gamma + 1}{2}\, \phi_x^2]\, d\tilde{y}_s - [\phi_{\tilde{y}}]\, dx_s = 0 \;, \tag{11a}$$

and by the condition that ϕ is continuous,

$$[\phi] = 0 \quad \text{or} \quad [\phi_x]\, dx_s + [\phi_{\tilde{y}}]\, d\tilde{y}_s = 0 \;. \tag{11b}$$

Here $(dx_s, d\tilde{y}_s)$ are line elements in the shock surface, and

$$[\;] = \text{jump} = (\;)_b - (\;)_a \quad,$$

where

$$(\;)_b = \text{quantity behind shock} \;,$$

$$(\;)_a = \text{quantity ahead of shock} \;.$$

Further, the mass flux in the x-direction is

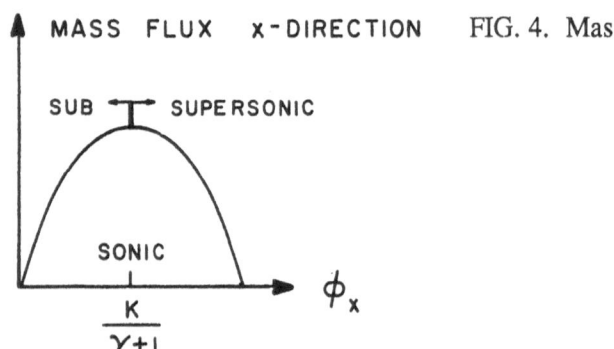

FIG. 4. Mass flux.

$$\frac{\rho q_x}{\rho_\infty U} = 1 + \delta^{4/3}(K\phi_x - \frac{\gamma+1}{2}\phi_x^2) + \cdots, \tag{12}$$

so that there is maximum flux at local sonic speed (cf. Fig. 4). This maximum in the mass flux corresponds to the fact that stream tubes have throats at local Mach number unity. The TSD equations thus contain all the essential features of the flow.

III. Early GALCIT experiments

A small 5-cm by 50-cm transonic wind tunnel was constructed in 1944-5 at GALCIT by Hans W. Liepmann, and a series of productive experiments were carried out in this facility[4]. Some of these are mentioned here and some in later sections. Experiments were carried out in flow past a series of circular-arc airfoils having dimensions as in Fig. 5. Surface-pressure distributions were recorded and schlieren pictures were taken. Figure 6 shows a typical pressure distribution at zero angle of attack. The local supersonic zone is shown clearly, as is the substantial difference between flows with laminar and turbulent boundary layers on the surface. Turbulence was induced with a trip wire. For the turbulent case, for which the boundary layer is thinner, the shock wave terminating the sonic region is very evident. These features are visible also in the schlieren photographs reproduced in Fig. 7.

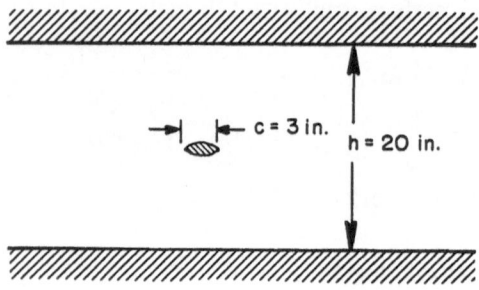

FIG. 5. Wind-tunnel geometry.

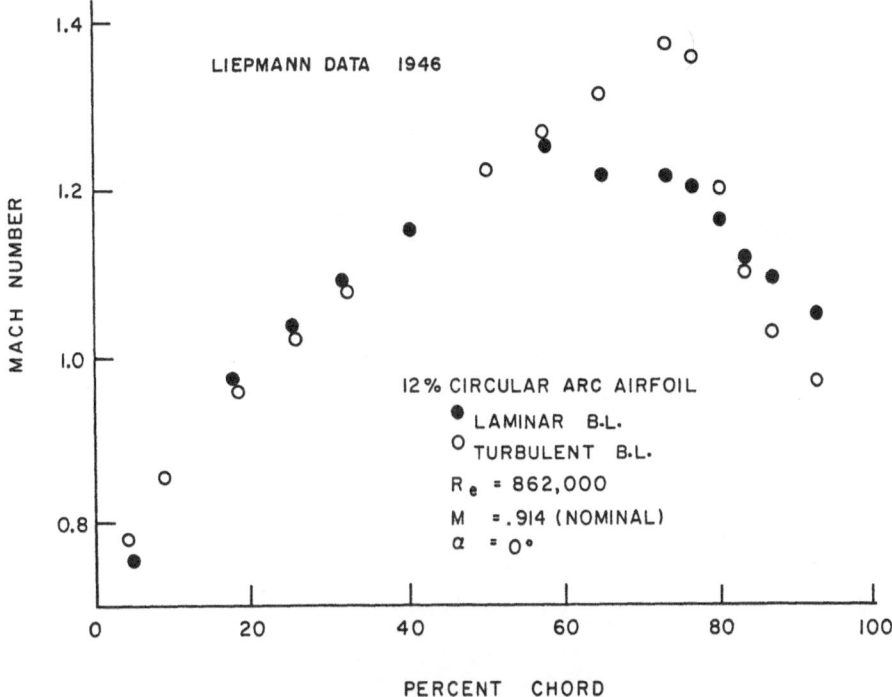

FIG. 6. Experimental pressure distribution -- circular arc[4].

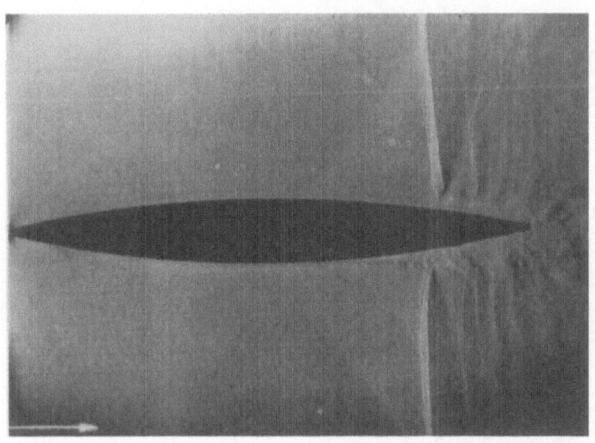

$M_\infty = 0.915$, $\alpha = 0°$, knife edge vertical, laminar boundary layer.

$M_\infty = 0.915;\quad \alpha = 0°$, knife edge horizontal, laminar boundary layer.

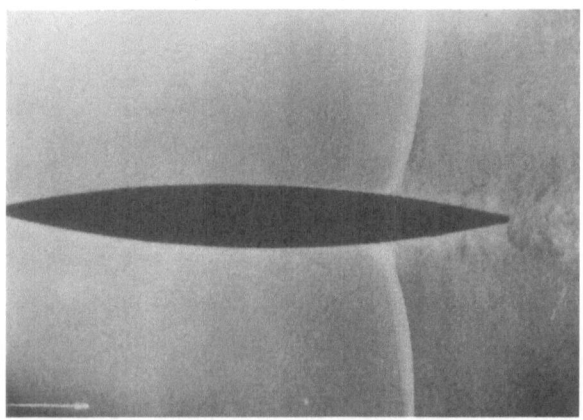

$M_\infty = 0.915,\quad \alpha = 0°$, knife edge vertical, turbulent boundary layer.

FIG. 7. Schlieren photographs of circular-arc airfoils[4].

$M_\infty = 0.915,\quad \alpha = 0°$, knife edge horizontal, turbulent boundary layer.

IV. Numerical methods

At the time of these experiments, no reliable numerical methods existed for calculating the ideal flow, although Emmons carried out some relaxation calculations for flows with local supersonic regions. Emmons's method was inherently unstable and did not resolve shock waves. In the late 1960's Emmons's method was revised[8] to eliminate these drawbacks, and since then the method has undergone substantial development. The use of multigrids has speeded convergence, and the basic ideas have been extended to the Euler equations. Here we describe the original idea and present the results of a few calculations.

Finite-difference methods are used to solve numerically the boundary-value problem for ϕ corresponding to flow past an airfoil. The best results are obtained by using the conservative form (8),

$$(K\phi_x - \frac{\gamma+1}{2} \phi_x^2)_x + (\phi_{\bar{y}})_{\bar{y}} = 0 \ , \tag{13}$$

and a corresponding conservative finite-difference form. The essential boundary conditions are (i) tangent flow at the airfoil surface;

$$\phi_{\bar{y}}(x, 0\pm) = F'_{u,l}(x), \qquad 0 < x < 1 \ , \tag{14}$$

where $y = \delta F_{u,l}(x)$ represents the upper or lower surface, respectively; (ii) vanishing of perturbations at infinity;

$$\phi_x, \ \phi_y \to 0, \qquad (x^2 + \bar{y}^2)^{1/2} \to \infty \ ; \tag{15}$$

and (iii) the Kutta condition that the flow leaves the trailing edge smoothly. In TSD this requirement is equivalent to zero pressure loading at the trailing edge, or (cf. (7))

$$\phi_x(1, 0+) = \phi_x(1, 0-) \ . \tag{16}$$

The boundary-value problem is shown in Fig. 8, where in addition it is noted that there is a jump in ϕ across the wake;

$$[\phi] = \phi(x, 0+) - \phi(x, 0-) = \bar{\Gamma} \ , \tag{17}$$

where $\bar{\Gamma} = \oint (u \ dx + v \ d\bar{y})$ is the circulation. It is also noted that an asymptotic far field exists which for subsonic flow has the form of a circulation,

$$\phi \to -\frac{\bar{\Gamma}\theta}{2\pi} + \cdots, \qquad \theta = \tan^{-1} \frac{K^{1/2}\bar{y}}{x} \ . \tag{18}$$

Finite-difference calculations are carried out on the (i, j) mesh indicated. A conservative form is derived by considering fluxes around a central box. This reasoning

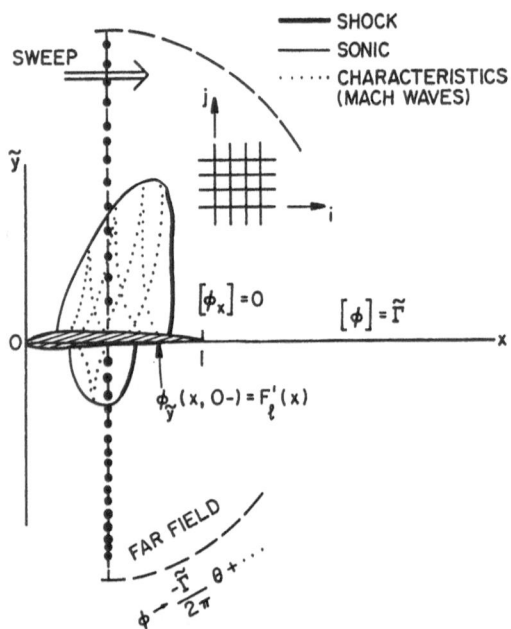

SHOCK
SONIC
CHARACTERISTICS (MACH WAVES)

SWEEP

$[\phi_x] = 0$

$[\phi] = \tilde{\Gamma}$

$\phi_y(x, 0-) = F'_\ell(x)$

FAR FIELD

$\phi \sim \dfrac{-\tilde{\Gamma}}{2\pi}\, \theta + \cdots$

FIG. 8. Boundary-value problem for calculations.

is extended so that the shock waves, which appear spread over three or four mesh points, are consistent. For stability, the difference scheme must be chosen to be type-sensitive. The solution at (i, j) can be influenced only by upstream points if the flow is locally supersonic $(\phi_x > K/(\gamma + 1))$, but by both upstream and downstream points if the flow is locally subsonic. In the finite-difference approximations, ϕ_x at (i, j) can be calculated from a centered formula $\phi_x^{(c)}$ involving $(i + 1, i - 1)$ or a backward formula $\phi_x^{(b)}$ involving $(i, i - 2)$. When these agree, an (i, j) can be designated subsonic or supersonic as indicated in Fig. 9. For subsonic points, (13) provides an explicit equation for ϕ_{ij} in terms of neighbors on all sides; see the computational star in Fig. 9. For supersonic points, however, an implicit scheme is used, involving only

$\phi_x^{(c)}$	$\phi_x^{(b)}$	TYPE	STAR
$<$	$<$	ELLIPTIC (SUB)	$(i-1) \bullet \; \boxed{x} \; \bullet \; (i+1)$
$>$	$>$	HYPERBOLIC (SUPER)	$\bullet \; \bullet \; \begin{smallmatrix} x \\ x \\ x \end{smallmatrix} (i, j)$
$>$	$<$	SONIC	$\begin{smallmatrix} x \\ x \\ x \end{smallmatrix}$
$<$	$>$	SHOCK PT. (STRONG)	$\bullet \; \bullet \; \begin{smallmatrix} x \\ x \\ x \end{smallmatrix} \; \bullet$

FIG. 9. Table of difference operators.

upstream points. Two other kinds of points, denoted as "sonic" for points near the sonic line where the flow accelerates, and "shock" for points where the flow decelerates to subsonic speed through a shock, are also shown.

Since the problem is non-linear, the local state is not known in advance, and an iteration scheme is used. At a given iteration the difference schemes are chosen and the unknowns solved for on a vertical line. Sweeps are made in the downstream direction. The analytical far field is used as a boundary condition, but the value of $\tilde{\Gamma}$ is adjusted as the trailing edge is passed. To speed convergence, the latest values are used whenever possible. The method converges well, gives the correct shock jumps, and for lifting cases automatically satisfies the Kutta condition.

Some results of calculations are shown in Figs. 10 and 11 for a parabolic-arc airfoil[9] and an NACA 0012 airfoil[10]. For the first case, the sonic line and shock are shown as

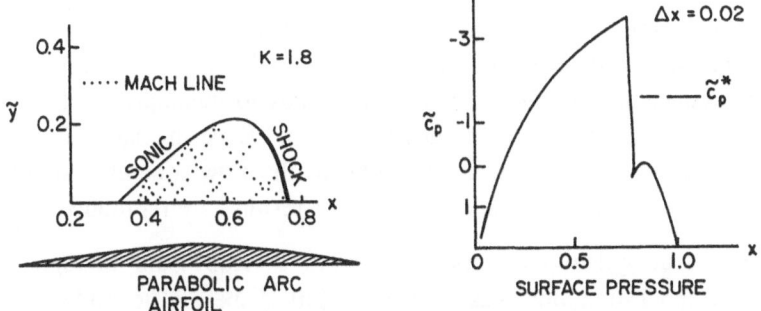

FIG. 10. Flow features and surface pressure -- parabolic arc.

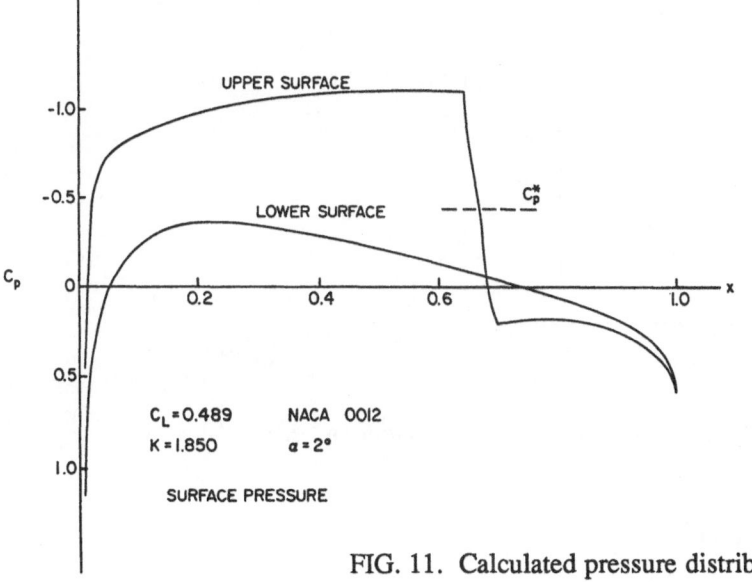

FIG. 11. Calculated pressure distribution[10].

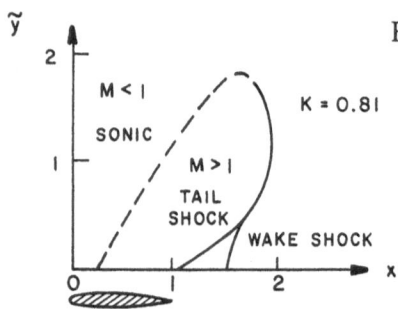

FIG. 12. Flow field features -- parabolic arc.

well as the surface pressure. Figure 12 shows the flow-field features[9] for a higher free-stream Mach number where the main shock has moved aft of the airfoil and a fishtail shock pattern appears.

V. Shock-free supercritical flows

The possibility of shock-free flows past airfoils with local supersonic regions was of considerable interest in the early days of transonic research. This interest was stimulated by Ringleb's exact hodograph solution[11] (analogous to incompressible flow around a half-plane), which has a smooth transonic region. Experimentally it was possible to produce a small shock-free supersonic zone in the flow around a simple shape. For example, see Fig. 13, where some of Liepmann's results[4] are reproduced.

Other exact solutions, analytical and numerical, were derived from hodograph considerations and gave shock-free flow past special airfoil shapes. The hodograph equations are linear, so that solutions can be obtained immediately and the airfoil shape found later. For the TSD system, the hodograph equations are obtained by a direct interchange of dependent and independent variables. Rewrite (8) as the system

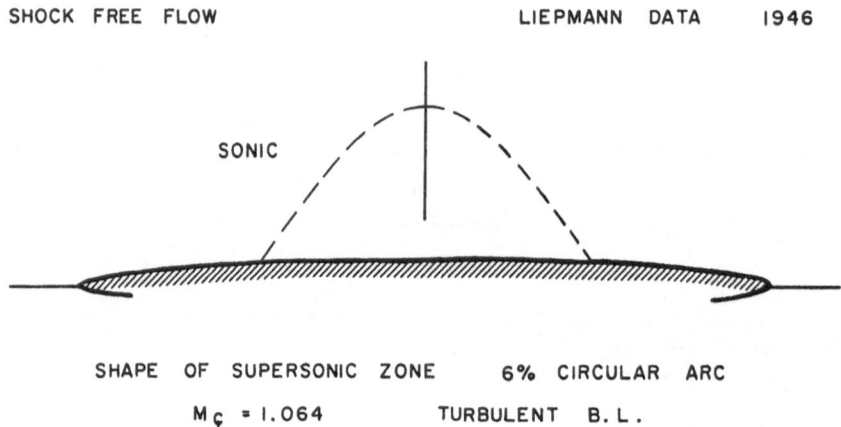

FIG. 13. Shock-free flow[4].

$$\begin{bmatrix} w \, w_x = v_{\tilde{y}} \\ w_{\tilde{y}} = v_x \end{bmatrix} \quad , \tag{19}$$

where

$$w = (\gamma + 1) \, \phi_x - K, \qquad v = (\gamma + 1) \, \phi_{\tilde{y}} \ .$$

Then

$$\begin{bmatrix} w \, \tilde{y}_v = x_w \\ x_v = \tilde{y}_w \end{bmatrix} \quad , \tag{20}$$

since $w_x = \tilde{y}_v / J$, etc., where $J = \text{Jacobian} = \partial(x, \tilde{y})/\partial(w, v)$. From this it follows that the approximate stream function $\tilde{y}(w, v)$ is a solution of Tricomi's equation,

$$w \, \tilde{y}_{vv} - \tilde{y}_{ww} = 0 \ . \tag{21}$$

This is the simplest linear equation of mixed type; elliptic in the subsonic region $w < 0$, hyperbolic in the supersonic region $w > 0$. The exact potential equation can be transformed in a similar way to produce Chaplygin's equation, which has properties analogous to (21).

Some special exact solutions for mixed flow past an airfoil were given by Tomotika and Tamada[12]. Recent advances have occurred in the work of Nieuwland[13], who used Chaplygin functions to represent families of airfoils analytically, and Garabedian and Korn[14], who used a finite-difference hodograph method to obtain numerical solutions for airfoil flows. The latter employ real characteristics in the hyperbolic region and complex characteristics in the elliptic region, using a sufficient number of parameters to generate families of shapes with local subsonic zones. A typical airfoil and its pressure distribution in shock-free flow are shown in Fig. 14. The pressure distribution is shown calculated according to TSD theory with a fully conservative relaxation scheme (FCR), as described in Sec. IV, and a non-conservative scheme[15] (NCR), together with the result of a calculation by Garabedian, Korn, and Jameson[16]. The drag coefficient, which is theoretically zero (see the next section) is evaluated on several different control surfaces[15].

The smooth hodograph solutions are seen to be isolated solutions, since a smooth mixed flow can no longer be found when the boundary conditions in the physical plane are changed slightly (Morawetz[17]). An early paper by Guderley[18] suggests a singularity of a perturbation in the downstream corner of the sonic region. These are isolated solutions in the same sense as for the Busemann supersonic biplane. When the conditions are changed slightly, a neighboring solution is found with a shock wave. An example of a calculation of such a flow appears in Fig. 15, where the shock wave is apparent. Experiments verify these features.

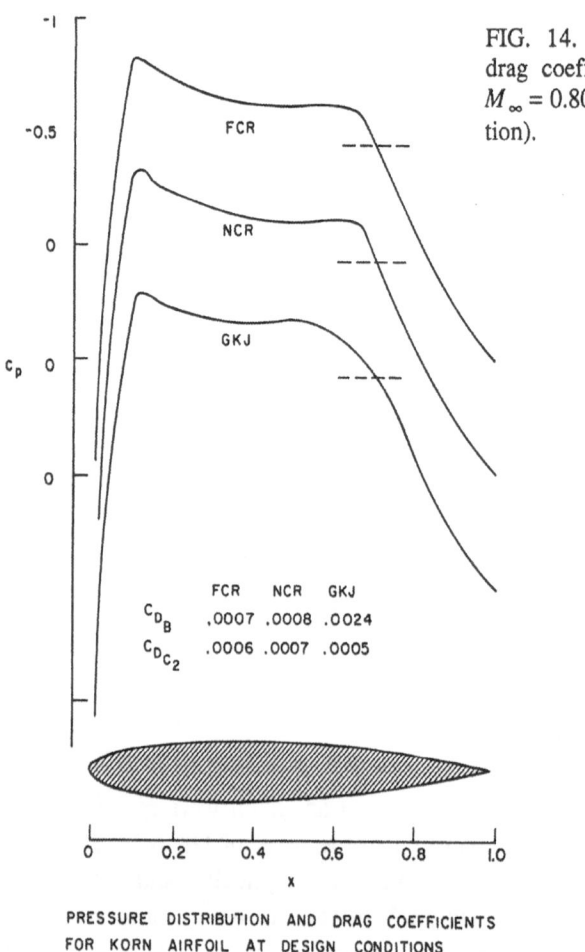

FIG. 14. Pressure distribution and drag coefficients for Korn airfoil at $M_\infty = 0.80$, $\alpha = 0°$ (design condition).

PRESSURE DISTRIBUTION AND DRAG COEFFICIENTS
FOR KORN AIRFOIL AT DESIGN CONDITIONS
$M_\infty = 0.80$, $\alpha = 0°$

Guderley[3] suggested that the shocks in the local supersonic region are initiated by an envelope formation of the compressive reflection of Mach lines from the sonic line. A detailed inspection of the calculations underlying Fig. 10 verifies this idea[9]. Thus a shock-free airfoil is one for which the shape of the sonic line is just right so that no envelope is formed.

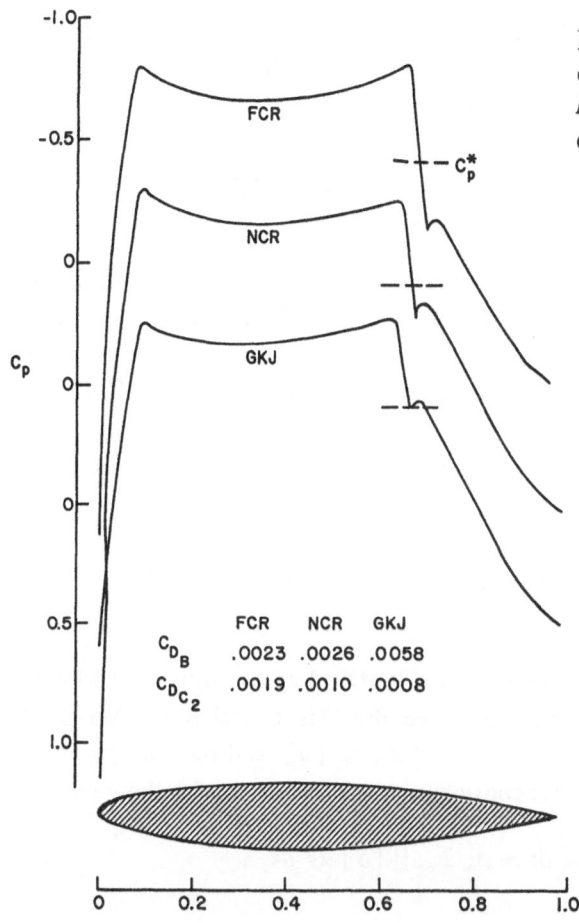

FIG. 15. Pressure distributions and drag coefficients for Korn airfoil at $M_\infty = 0.81$, $\alpha = 0°$ (off-design condition).

PRESSURE DISTRIBUTION AND DRAG COEFFICIENTS
FOR KORN AIRFOIL AT OFF DESIGN CONDITION
$M_\infty = 0.81$, $\alpha = 0°$

VI. Shock waves and drag

The connection between shock waves and drag was made explicit by Liepmann[19] in 1950 in a study of linearized supersonic flow past an airfoil. By considering weak shock waves in such a flow, a correction was made to the Mach angle for shock angle and location, and the formula was found

$$\text{drag}/\text{length} = \rho_\infty T_\infty \int_{\text{shocks}} [S]\, dy \ , \tag{22}$$

where $[S]$ = jump in specific entropy across a shock wave. Drag is directly related to entropy production. Similar results appear later in works by von Karman[20] and Oswatitsch[21].

Analogous considerations apply to TSD flow. Germain[22] gave a derivation of a drag formula for TSD theory starting from a local conservation law in two dimensions. The generalization of this law to three-dimensional flow[5] reads

$$\left[K \frac{u^2}{2} - \frac{v^2 + w^2}{2} - \frac{\gamma + 1}{3} u^3 \right]_x + (uv)_{\bar{y}} + (uw)_{\bar{z}} = 0 . \tag{23}$$

This conservation law follows easily from the three-dimensional version of (8) and the equation of irrotationality. The values of (uv) on $\bar{y} = 0$ are proportional to the incremental drag for a planar system, since the pressure increment is proportional to u and the airfoil slope to v. Integration of (23) over a control surface, as shown in Fig. 16, yields a formula for drag. Since (23) is not a physical global conservation law, it is not valid across shock waves. As the control surface grows to infinity, contributions from the shock and vortex sheet remain. The final result is

$$\bar{D} = \frac{drag}{\rho_\infty U^2 \delta^{4/3} c^2} = \lim_{x \to \infty} \iint\limits_{-\infty}^{\infty} (\frac{v^2 + w^2}{2}) \, d\bar{y} \, d\bar{z} - \frac{\gamma + 1}{12} \iint [u]^3 \, d\bar{y} \, d\bar{z} . \tag{24}$$

The first term of (24) is the familiar expression for vortex drag in terms of the kinetic energy in the wake. The second term, the wave drag, is a scaled version of the Liepmann formula. The jump of ϕ_x across the shock is $[u]$ and the entropy change is proportional to $[u]^3$. Thermodynamic considerations for weak shocks show that

$$\frac{\gamma + 1}{12} \iint [u]^3 \, d\bar{y} \, d\bar{z} \sim \rho_\infty T_\infty \iint [S] \, dy \, dz . \tag{25}$$

It should be noted that these simple results do not carry over for stronger shocks.

The drag-entropy formula (24) can be used to provide a check on the consistency of numerical TSD calculations[15]. In Fig. (17), the shock and flow field for a parabolic-arc airfoil at $M_\infty = 0.909$ are shown. The quantity $\int [u]^3 d\bar{y} \sim \int [c_p]^3 d\bar{y}$ is plotted also, and the drag coefficient is computed both from the surface pressure integral and the entropy jump. The result is

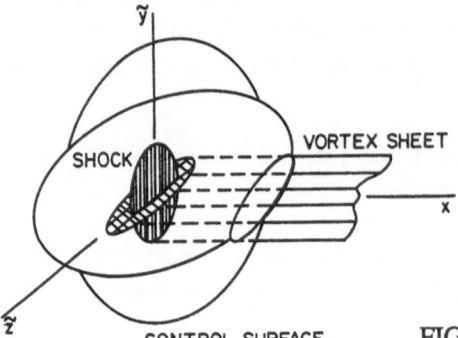

FIG. 16. Control volume for drag.

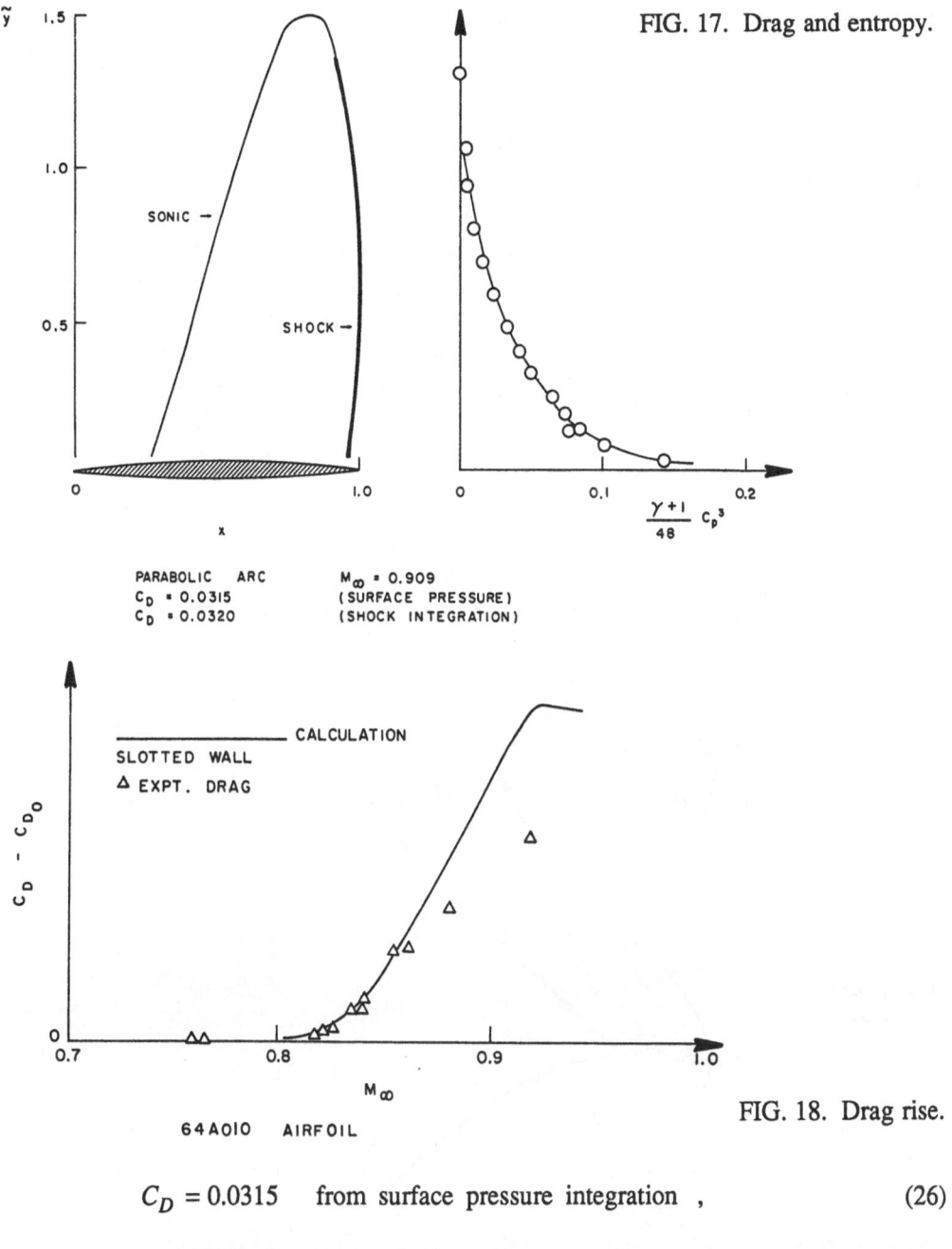

FIG. 17. Drag and entropy.

PARABOLIC ARC
C_D = 0.0315
C_D = 0.0320

M_∞ = 0.909
(SURFACE PRESSURE)
(SHOCK INTEGRATION)

FIG. 18. Drag rise.

$$C_D = 0.0315 \quad \text{from surface pressure integration} , \tag{26}$$

$$C_D = 0.0320 \quad \text{from shock entropy integration} .$$

Careful calculations of TSD flow enable the wave drag to be found. An example of an application is the flow past an NACA 64A010 airfoil in a slotted-wall wind tunnel; see Fig. 18. The drag rise due to shock-wave formation (C_{D_0} = friction drag) is fairly well represented.

VII. Sonic flow and the law of stabilization

The special structure of flow at $M_\infty = 1$ was first elucidated by Guderley and Frankl. In order to understand this flow, consider the sequence of flow patterns past an airfoil or body at free-stream Mach numbers close to one, as shown in Fig. 19. At high subsonic Mach numbers, a large supersonic zone is terminated by an oblique shock near the trailing edge and by the fishtail shock, as calculated in Fig. 12. At $M_\infty = 1$, the supersonic zone in the sequence of steady flows grows to infinity and the fishtail shock moves to downstream infinity. A limit characteristic or Mach wave appears which at infinity is asymptotically parallel to both the sonic line and the tail shock. This limit characteristic divides the flow field into an upstream and a downstream part. Any (infinitesimal) disturbance to the flow in the supersonic region, for example, can send a disturbance downstream which eventually reaches the sonic line and affects the entire subsonic region. Any disturbance originating downstream of the limit characteristic

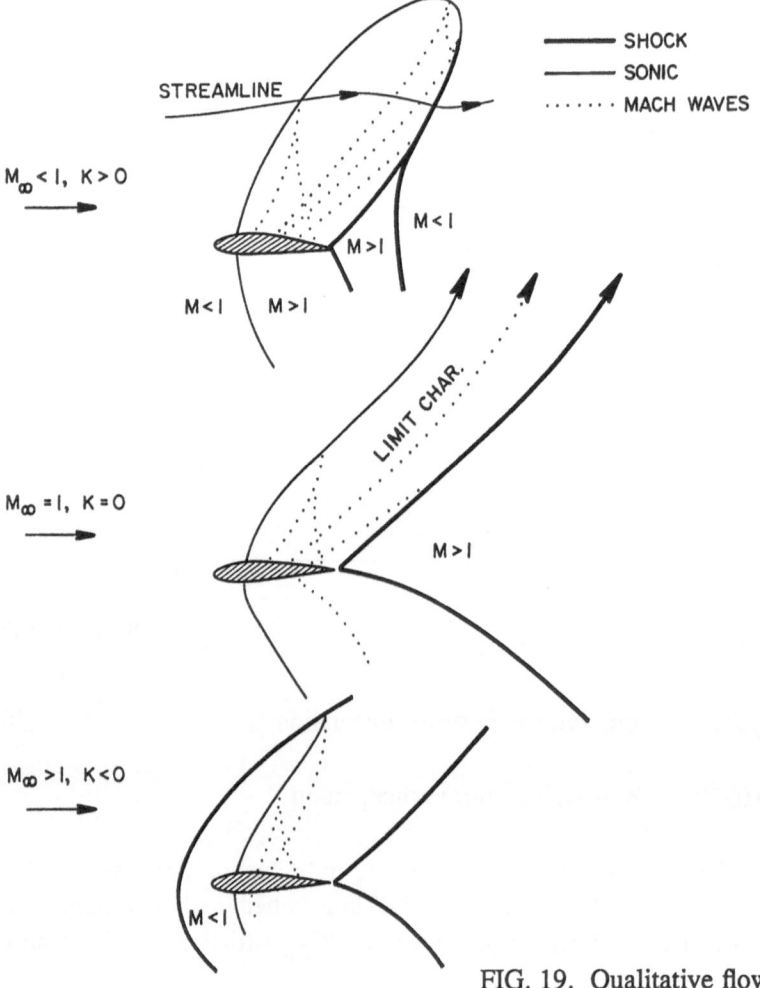

FIG. 19. Qualitative flow structure, $M_\infty \approx 1$.

cannot affect the upstream flow. The flow in the upstream section, up to the limit characteristic, thus has an elliptic nature and must be calculated all at once. It is effectively independent of the flow downstream of this characteristic. Downstream of the limit characteristic the flow can, for example, be calculated as in a hyperbolic region by the method of characteristics. When the free stream becomes slightly supersonic a detached shock wave, with subsonic flow behind it near the axis $(\bar{y} = 0)$, appears ahead of the body. The flow becomes supersonic near the airfoil and terminates again in an oblique tail shock. It can thus be appreciated that the flow in the neighborhood of the body does not change qualitatively very much, since the oncoming flow is always close to a uniform subsonic flow.

For $M_\infty = 1$, the far field is a similarity solution. With $K = 0$, (8) becomes, for the planar case,

$$(\gamma + 1) \, \phi_x \phi_{xx} - \phi_{\bar{y}\bar{y}} = 0 \ . \tag{27}$$

The far field can be thought of as being produced by a singularity at the origin. In view of the non-linearity, the far field can be represented in the form

$$(\gamma + 1) \, \phi(x, \bar{y}) = \bar{y}^{3\kappa - 2} f(\xi; \kappa), \qquad \xi = \frac{x}{\bar{y}^\kappa} \ , \tag{28}$$

which gives a one-parameter family of smilarity solutions. Thus

$$(\gamma + 1) \, \phi_x = \bar{y}^{2\kappa - 2} f'(\xi), \qquad (\gamma + 1) \, \phi_{xx} = \bar{y}^{\kappa - 2} f''(\xi) \ , \tag{29}$$

where $f(\xi; \kappa)$ is found from

$$(f' - \kappa^2 \xi^2) \, f'' - 5\kappa (1 - \kappa) \xi f' + 3(1 - \kappa)(3\kappa - 2) f = 0, \ -\infty < \xi < \xi_L \tag{30}$$

The coordinates are illustrated in Fig. 20. The flow starts at infinity, decelerates and spreads around the body, and then accelerates smoothly through sonic speed and through the limit characteristic. It can be seen from (29) that the sonic line $(f' = 0)$,

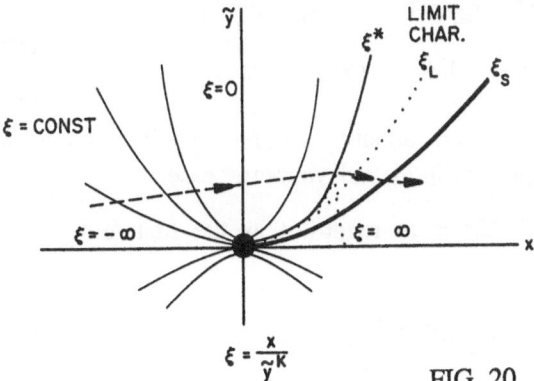

FIG. 20. Similarity curves.

as well as the limit characteristic and the tail shock, must lie on similarity curves ξ, ξ_L, and ξ_s, respectively. The characteristic condition (29) becomes (+)

$$f'(\xi_L) = \kappa^2 \xi_L^2 . \tag{31}$$

Also, ϕ must be symmetric in $\tilde{y}$, since the apparent-thickness effect dominates the flow at infinity. This occurs because both acceleration and deceleration of a sonic flow produce a widening of stream tubes (cf. Fig. 4). One constant of integration remains if integration is started at $\xi = -\infty$, and two conditions must be satisfied at the limit characteristic if $f''(\xi_L)$ is to be finite. Thus the solution exists only for a special value of κ. Guderley showed from the hodograph solution that

$$\kappa = 4/5 \quad \text{for the planar case} \tag{32}$$

and by numerical integration that $\kappa = 4/7$ for the analogous axisymmetric case. The latter result was later shown to be exact by various writers[23]. Equation (30) has a group property, so that the solution can be written

$$f(\xi) = \frac{1}{A^3} F(A\xi), \qquad A = \text{constant} . \tag{33}$$

The scale factor A depends on the size and shape of the body. Thus the far field can be standardized to have $\xi_L = 1$. By considering shock jumps, the solution can be continued to downstream infinity. A particularly useful representation of the solution was given parametrically by Frankl[24];

$$f(s) = a_1^{3/5} s^{-1/5} \left[1 - \frac{s}{2} + \frac{s^3}{3} \right], \qquad a_1 = \frac{2^9 \, 3^3}{5^5} , \tag{34}$$

$$\xi(s) = a_1^{1/5} s^{-2/5} (s - \frac{1}{2}) ,$$

where the following relationships are noted;

ξ	$-\infty$	0	ξ^*	$\xi_L = 1$	ξ_s
s	0	1/2	1	4/3	$(5\sqrt{3} + 8)/6$

(35)

Frankl produced this result by clever observations about special hodograph solutions; an analogous result has been derived by inspection in the axisymmetric case.

A useful extension of the far field (28) is to regard it as the first term of an expansion of the form

$$(\gamma + 1) \, \phi(x, \tilde{y}) = \tilde{y}^{2/5} \frac{1}{A^3} f(A\xi) + C_0 \tilde{y}^{\sigma_0} \frac{f_0(A\xi)}{A^3} + C_1 \tilde{y}^{\sigma_1} \frac{f_1(A\xi)}{A^3} . \tag{36}$$

This expansion is valid as $\bar{y} \to \infty$ for ξ fixed. The perturbation functions $f_i = g$, $i = 0,\ 1$, satisfy the variational equation (with $\sigma_i = \alpha$),

$$(f' - \frac{16}{25}\xi^2)\, g'' + \left[f'' + \frac{4}{5}(2\alpha - \frac{9}{5})\xi \right] g' - \alpha\,(\alpha - 1)\, g = 0 \ . \tag{37}$$

If $g = s^{-\alpha/2}\, h(t)$, $t = 3s/4$, (37) becomes a standard hypergeometric equation,

$$t\,(1 - t)\, h'' + \left[\frac{1}{2} - \frac{1}{3}(5\alpha + 4)\, t \right] h' + \frac{5}{12}\,\alpha\,(5\alpha - 2)\, h = 0 \ , \tag{38}$$

in the interval $-\infty \le \xi \le \xi_L = 1$, $0 \le t \le 1$. The two linearly independent solutions around $t = 0$ are

$$h_I = F(\frac{5}{2}\alpha,\ \frac{1}{6}(2 - 5\alpha);\ \frac{1}{2};\, t) \ , \tag{39}$$

$$h_{II} = t^{1/2}\, F(\frac{1}{2}(5\alpha + 1),\ \frac{5}{6}(1 - \alpha);\ \frac{3}{2};\, t) \ .$$

Thus solutions that are smooth on passing throught the limit characteristic are given by the spectrum

$$\sigma_i: \quad \cdots \quad -\frac{3}{5},\, -\frac{2}{5},\, -\frac{1}{5},\, 0,\, \frac{2}{5},\, \frac{8}{5},\, \frac{11}{5},\, \frac{14}{5} \quad \cdots \tag{40}$$

For the expansion near infinity, $\sigma_0 = 0$, $\sigma_1 = -1/5$, and the f_i are simple functions.

The solutions just discussed can be used to relate a flow at a free-stream Mach number close to unity to a flow at Mach number unity. In TSD form, this relates a flow with small $|K|$ to a flow with $K = 0$. For flow that is subsonic at infinity, the dominant term in the far field is the circulation term, basically the solution that decays most slowly for

$$K\phi_{xx} + \phi_{yy} = 0, \qquad \phi \to -\frac{\Gamma}{2\pi}\theta \ . \tag{41}$$

But for $K = 0$ the far field is given by (26) with $\kappa = 4/5$. Thus there is a non-uniformity at infinity in $\phi\,(x, \bar{y}; K)$ as $K \to 0$, and inner and outer expansions are needed. For the inner expansion, valid near the airfoil, the first term is sonic flow past the airfoil, and corrections are sought. In early work, Liepmann and Bryson[25] proposed that near $M_\infty = 1$ the local Mach number on a body does not change as the free-stream Mach number changes, because of the qualitative ideas outlined above. The Law of Stabilization proposed by Ryzhov and Lifschitz[26], and the more detailed work by Cook and Ziegler[27] using the method of matched asymptotic expansions[28], give a deeper and more precise description of this Mach-number freeze. The starting point is a solution at $M_\infty = 1$. Several are available; for example, Guderley and Yoshihara[3] gave the result for a wedge, while Tse[29] worked out lifting airfoils at $M_\infty = 1$. Both of these methods rely on a hodograph formulation. Let $\phi^*(x, \tilde{y})$ represent the sonic flow past an airfoil with the far field (28). Then the flow for $K > 0$ is represented as an inner expansion, generated from a limit process $K \to 0$, x, $\tilde{y}$ fixed,

$$\phi(x, \tilde{y}; K) = \phi^*(x, \tilde{y}) + K \frac{x}{\gamma + 1} + \varepsilon(K) \phi_c(x, \tilde{y}) + \cdots \quad . \tag{42}$$

Here $\varepsilon(K)$ is the order of magnitude of the correction that is sought, and $\phi_c(x, \tilde{y})$ is the correction potential. The quantity $\phi^*(x, \tilde{y})$ satisfies the sonic TSD equation (25) and its boundary conditions, while $\phi_c(x, \tilde{y})$ is found from the variational equation

$$(\gamma + 1)(\phi_x^* \phi_{c_{xx}} + \phi_{xx}^* \phi_{c_x}) - \phi_{c_{\tilde{y}\tilde{y}}} = 0 \quad . \tag{43}$$

The expansion (42) is not valid near infinity, so that an outer expansion in rescaled variables is sought;

$$\phi(x, \tilde{y}; K) = \sigma(K) \bar{\phi}(\bar{x}, \bar{y}) + \cdots \quad . \tag{44}$$

The associated limit process has $\bar{x}, \bar{y}$ fixed as $K \to 0$, $x, \tilde{y} \to \infty$; that is,

$$\bar{x} = \mu(K) x, \qquad \bar{y} = \nu(K) \tilde{y}, \qquad \mu, \nu \to 0 \quad .$$

The condition that the rescaled equation be subsonic at infinity but still nonlinear and of changing type produces a one-parameter family of flows,

$$\phi(x, \tilde{y}; K) = \frac{K}{\mu(K)} \bar{\phi}(\bar{x}, \bar{y}) + \cdots, \quad \text{with } \nu = \mu K^{1/2} \quad , \tag{45}$$

with the resulting version of the TSD equation,

$$\left[1 - (\gamma + 1) \bar{\phi}_x\right] \bar{\phi}_{xx} + \bar{\phi}_{yy} = 0 \quad . \tag{46}$$

The inner and outer expansions must match in a simple way as $x, \tilde{y} \to \infty$, $\bar{x}, \bar{y} \to 0$. Now ϕ^* is defined along similarity curves which then must match;

$$\xi = \frac{x}{\tilde{y}^{4/5}} = \frac{K^{2/5}}{\mu^{1/5}} \frac{\bar{x}}{\bar{y}^{4/5}} = \bar{\xi} \quad .$$

Thus $\mu = K^2$, $\bar{x} = K^2 x$, $\bar{y} = K^{5/2} \tilde{y}$. The general appearance of the flow in outer variables is shown in Fig. 21. Higher-order terms in the flow near infinity for the inner expansion (42) and the flow near the origin in the outer expansion (45) then depend on the spectrum associated with (37).

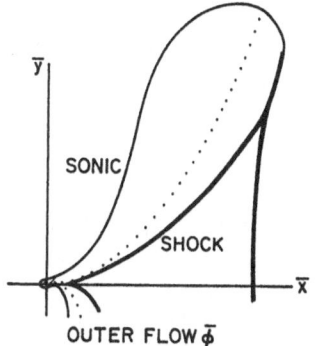

SONIC

SHOCK

OUTER FLOW $\bar{\phi}$

FIG. 21. Outer flow -- law of stabilization.

These expansions can be written (with the scale factor $A = 1$)

$$\phi(x, \bar{y}; K) = \frac{1}{\gamma + 1} \left[\bar{y}^{2/5} f(\xi) + c_0 + \bar{y}^{-1/5} f_1(\xi) + \cdots \right] \tag{47a}$$

$$+ \frac{K x}{\gamma + 1} + \varepsilon(K) \phi_c(x, \bar{y}) + \cdots$$

$$= \frac{1}{K} \frac{1}{\gamma + 1} \left[\bar{y}^{2/5} f(\bar{\xi}) + \bar{x} + C_1 \bar{y}^{8/5} \bar{f}_1(\bar{\xi}) + \cdots \right]. \tag{47b}$$

A comparison of these two expressions as $\bar{x}, \bar{y} \to 0$, $x, \bar{y} \to \infty$ along similarity curves shows that $\phi_1(x, \bar{y})$ has the similarity form at infinity and that

$$\varepsilon(K) = K^3 = \frac{(1 - M_\infty^2)^3}{\delta^2}. \tag{48}$$

This weak dependence of the potential and the pressure distribution on the deviation from $M_\infty = 1$ is the essential part of the Law of Stabilization. The term $K x/(\gamma + 1)$ in these expansions represents the change of the flow at infinity and adds a constant pressure level to the solution. Germain[22] derived a conservation-law formula for the scale factor A, relating it to properties of the solution (not known in advance) on the body surface. Cook and Ziegler[27] have extended these ideas to find C_1. The boundary-value problem for the inner correction solution is sketched in Fig. 22. An empirical fit of the Law of Stabilization to some experimental drag measurements by Vincenti[30] is shown in Fig. 23.

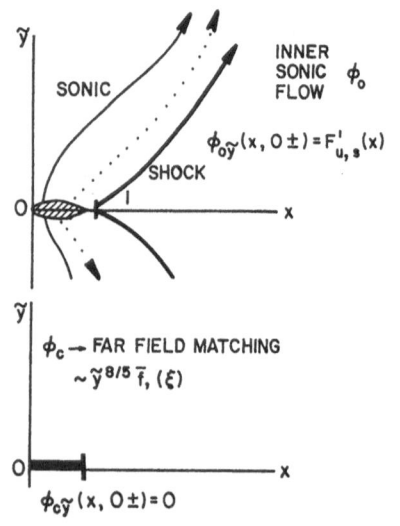

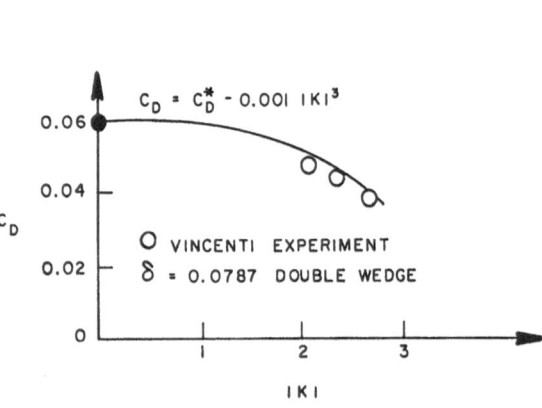

FIG. 22. Boundary-value problem for ϕ_c.

FIG. 23. Law of stabilization (data from Ref. 30).

VIII. Concluding remarks

TSD theory and calculations have made great advances in the last 35 years, following the work of the pioneers in the field. The understanding of the physical phenomena that has been thus achieved enables us to go forward with more elaborate calculations and to include more physical effects. We can only admire the deep insight of H.W. Liepmann and other early workers in this area.

Acknowledgment

This research was sponsored by the Air Force Office of Scientific Research, Air Force Systems Command, USAF, under Contract/Grant No. AFOSR-82-0155. The United States Government is authorized to reproduce and distribute reprints for governmental purposes notwithstanding any copyright notation thereon.

References

1. S.A. Chaplygin, *On gas jets* (1902), in *Collected Works*, Vol. II (Government Press, Moscow, 1948; in Russian). English translation, NACA TM 1063, 1944.

2. T. von Karman, "The similarity law of transonic flow," J. Math. Phys. **26**, 182-190 (1947).

3. K.G. Guderley, *Theorie schallnaher Strömungen* (Springer-Verlag, Berlin, 1957). English translation, *The Theory of Transonic Flow* (Pergamon Press, 1962).

4. H.W. Liepmann, H. Ashkenas, and J.D. Cole, "Experiments in transonic flow," Wright-Patterson AFB, USAF TR 5667, 1948.

5. J.D. Cole, "Twenty years of transonic flow," Boeing Scientific Res. Labs., Rep. D1-82-0878, 1969.

6. R. Timman, "Unsteady motion in transonic flow," in *Symposium Transsonicum*, edited by K. Oswatitsch (Springer-Verlag, Aachen, 1962), pp. 394-401.

7. J.A. Krupp and J.D. Cole, "Unsteady transonic flow," Studies in Transonic Flow IV, UCLA Eng. Rep. 76104, 1976.

8. E.M. Murman and J.D. Cole, "Calculation of plane steady transonic flows," AIAA J. **9**, 121-141 (1971).

9. E.M. Murman, "Analysis of embedded shock waves calculated by relaxation methods," AIAA J. **12**, 626-632 (1974).

10. R.D. Small, "Numerical solutions for transonic flows, Part 1. Plane steady flow over lifting airfoils," Dept. Aero. Eng., Technion, Haifa, TAE Rep. 273, 1976.

11. F. Ringleb, "Exakte Lösungen der Differentialgleichungen einer adiabatischen Gasströmung," Z. angew. Math. Mech. **20**, 185-198 (1940).

12. S. Tomotika and K. Tamada, "Studies on two dimensional transonic flows of compressible fluids, Parts I, II, III," Quart. Appl. Math. **7**, 381-397 (1950); **8**, 127-136 (1950); **9**, 129-147 (1951).

13. G.Y. Nieuwland, "Transonic potential flow around a family of quasi-elliptical aerofoil sections," Amsterdam, NLR Tech. Rep. R172, 1967.

14. P. Garabedian and D. Korn, "Numerical design of transonic airfoils," in *Numerical Solutions of Partial Differential Equations II* (Academic Press, 1971).

15. E.M. Murman and J.D. Cole, "Inviscid drag at transonic speeds," Studies in Transonic Flow III, UCLA-ENG-7603, 1975.

16. P.R. Garabedian and D.G. Korn, "Analysis of transonic airfoils," Comm. Pure Appl. Math. **24**, 841-851 (1971) (see also A. Jameson, Grumman Aerospace Corporation, Report 390-71-1, 1971).

17. C. Morawetz, "On the non-existence of continuous transonic flows past profiles," Comm. Pure Appl. Math. **9**,5-68 (1956); **10**, 107-131 (1957).

18. G. Guderley, "On the presence of shocks in mixed subsonic-supersonic flow patterns," Adv. Appl. Mech. **3**, 145-184 (1953).

19. H.W. Liepmann, "On the relation between wave drag and entropy increase," Douglas Aircraft Company, Rep. SM13726, 1950.

20. T. von Karman, "On the foundation of high speed aerodynamics," in *High Speed Aerodynamics and Jet Propulsion*, Vol. VI, edited by W.R. Sears (Princeton Univ. Press, 1957), pp. 3-30.

21. K. Oswatitsch, *Gas Dynamics* (Academic Press, 1956).

22. P. Germain, "Écoulements transsoniques homogènes," ONERA T.P. 242 (1965) (see also Prog. Aero. Sci. **5**, Pergamon Press, 1964).

23. E.A. Müller and K. Matschat, "Ähnlichkeitslösungen der transsonischen Gleichungen bei der Anström-Machzahl 1," *Proc. 11th Int'l. Cong. Appl. Mech.*, edited by H. Görtler (Springer-Verlag, 1966), pp. 1061-1068.

24. L.D. Landau and E.M. Lifshitz, *Fluid Mechanics* (English edition) (Pergamon Press, 1959).

25. H.W. Liepmann and A.E. Bryson, Jr., "Transonic flow past wedge sections," J. Aeron. Sci. **17**, 745-755 (1950).

26. Yu. B. Lifschitz and O. Ryzhov, "Transonic flow around a carrying airfoil profile," Akad. Nauk SSSR, Mekh. Zhidk. Gaza **1978**, No. 1, 104-112 (Fluid Dyn. **13**, 78-84 (1978)).

27. L.P. Cook and F. Ziegler, "The stabilization law for transonic flow," S.I.A.M. J. Appl. Math. **46**, 27-48 (1986).

28. J.D. Cole, "Asymptotic problems in transonic flow," AIAA paper 82-0104, 1982.

29. E. Tse, "Airfoils at sonic velocity," Ph.D. Thesis, Univ. of California (Los Angeles), 1981.

30. W.G. Vincenti, D.W. Dugan, and E.R. Phelps, "An experimental study of the lift and pressure distribution on a double wedge profile at Mach number near shock attachment," NACA TN 3225, 1954.

Perspectives in Vortex Dynamics

P.G. Saffman
Professor of Applied Mathematics
California Institute of Technology, Pasadena, California 91125

I. Introduction

Interest in vortices and whirlpools goes back to prehistory, and the vortex is a fairly common old religious symbol. In more recent times, philosophers and artists have been attracted by vortex motion. Beautiful drawings of vortices in turbulent flow were made by Leonardo da Vinci at the beginning of the 16th century. Van Gogh's painting "Starry Night" suggests that he was familiar with the phenomenon of vortex pairing. For further examples, see Lugt[1].

The mathematical study of vortex dynamics starts with Helmholtz's great paper of 1858 and continues with the work of Lord Kelvin and other 19th-century scientists. Vorticity is well described, in Küchemann's phrase, as "the sinews and muscles of fluid motion." Vorticity is clearly central to the motion of a relatively inviscid, incompressible fluid and is always being studied; but it appears to become particularly fashionable every 50 years or so, and we now seem to be in such a phase. We can list four reasons for the interest:

1. Vortex dynamics is a promising approach to turbulence.

2. It is crucial to the understanding of separated flow and aerodynamic control.

3. It provides a physical example of a strongly non-linear, infinite-dimensional Hamiltonian system (at least when $v = 0$), and is therefore of interest in connection with modern work on dynamical systems and chaotic phenomena.

4. It poses the intellectual challenge of difficult mathematical problems that stretch the available analytical and numerical resources.

Although all real flows must contain some vorticity, vortex dynamics is generally understood to mean the creation, structure, stability, interaction and evolution of regions of concentrated vorticity. It is a consequence of the Helmholtz laws of vortex motion that the equations for the evolution of the vorticity are closed, and it is sufficient to take as unknown variable the vorticity itself. In a two-dimensional inviscid flow with velocity vector $q = (u[x, y, t], v[x, y, t], 0)$ and vorticity vector $\omega = (0, 0, \omega[x, y, t])$, the evolution equation is the simplest possible differential equation,

$$\frac{d\omega}{dt} = 0 , \tag{1}$$

which says that vorticity is conserved following the fluid. In three dimensions, the evolution equation is not so simple, having the form

$$\frac{d\omega}{dt} = (\omega \cdot \nabla) \, \mathrm{curl}^{-1} \, \omega , \tag{2}$$

where the right-hand side describes the amplification of vorticity due to stretching of vortex lines. Note that the right-hand side can be written in tensor notation in the form $e_{ij} \omega_j$, where e_{ij} is the rate-of-strain tensor and is independent of the value of ω at the point. Thus, vorticity does not actually stretch itself. This fact is responsible for major difficulties in estimating the growth of vorticity and in determining if the three-dimensional equations of hydrodynamics are well posed.

Of course, these equations only give the magnitude of the vorticity. To determine its position relative to a fixed frame, it is necessary to replace the material time derivative d/dt by the Eulerian derivative $\partial/\partial t$ through the relation

$$\frac{d}{dt} = \frac{\partial}{\partial t} + \mathrm{curl}^{-1} \, \omega \cdot \nabla . \tag{3}$$

We write the velocity as $\mathrm{curl}^{-1} \, \omega$ to emphasize that the velocity field is determined by the vorticity distribution. The actual evaluation of the inverse curl operator involves integrating over the vorticity and finding additional irrotational velocity fields needed to satisfy boundary conditions. In any case, the two-dimensional situation is now not quite so simple. But two dimensions are still orders of magnitude simpler than three dimensions, and the main body of work on vortex dynamics is restricted to two dimensions. Phenomena such as the Karman vortex street and the quasi-two-dimensional structures of the turbulent mixing layer provide some justification for this initial (100 year!) interest. Moreover, slowly varying, three-dimensional flows such as vortex-sheet roll-up and trailing-vortex formation behind lifting bodies can be approximated as evolving two-dimensional flows.

II. Basic problems in two dimensions

Vortex motions in two dimensions can be conveniently separated into problems involving the motion of points, sheets and patches. We discuss these in turn.

Points. This case is a simple representation in which the vorticity is described by a collection (finite and enumerable) of delta functions,

$$\omega = \sum_j \Gamma_j \, \delta(r - r_j(t)) , \tag{4}$$

where $r = (x, y)$. Each vortex then moves under the velocity induced by all the others;

$$dz_k/dt = (i/2\pi) \sum'_j \Gamma_j/(z_k - z_j)^* ,\qquad (5)$$

where $z_k = x_k + iy_k$. The prime means that $j = k$ is omitted from the sum, and $*$ denotes complex conjugation. If appropriate, irrotational velocity fields produced by boundaries are added to the right-hand side of (5). In this way, we have reduced the problem of fluid motion to a set (possibly enumerably infinite) of non-linear, coupled ordinary differential equations. Note, incidentally, that we are dealing not with just a numerical approximation but with a weak solution of the Euler equations.

This problem is perfect for computation with modern digital computers, and a considerable amount of electricity is being consumed computing solutions of (5) and related equations. In fact, the most popular way to investigate the motion of sheets and patches is to replace them by a large number of point vortices. The validity of this approach is an important unresolved question in numerical analysis.

Further, the system (5) is a Hamiltonian system with x_j and y_j as canonical conjugate variables. This statement remains true in the presence of solid boundaries[2]. The motion of vortex points can therefore provide pretty examples of the behavior of dynamical systems having few degrees of freedom, as well as problems in statistical mechanics in which the number of vortices is very large (these problems are still essentially unsolved, or at least the solutions have not been presented in easily understandable form).

Sheets. A vortex sheet is a surface (in two dimensions, a curve) across which the tangential velocity is discontinuous, or equivalently a surface on which the vorticity is infinite with a finite integral. An extension of (5) can be given to describe the motion of a two-dimensional sheet. Let $Z(\Gamma, t)$ be the parametrization of the equation of the curve, where Γ is the circulation between Z and some origin on the curve. Then the evolution of Z is described by the Birkhoff-Rott partial singular integro-differential equation,

$$\frac{\partial Z(\Gamma, t)}{\partial t} = \frac{i}{2\pi} P\!\int \frac{d\Gamma'}{[Z(\Gamma) - Z(\Gamma')]^*} ,\qquad (6)$$

where P denotes the Cauchy principal value. If appropriate, a further irrotational velocity field can be added to the right hand side of (6) to represent the effects of walls or other vortices.

The evidence is now extremely convincing that initial-value problems for the evolution of vortex sheets (the simplest is finite-amplitude Helmholtz instability of a straight uniform vortex sheet) are ill-posed; i.e., an initially smooth curve whose motion is described by (6) will develop a singularity (probably a cusp) in a finite time t_c , say. An excellent review of this hard problem (which still awaits rigorous solution) has been given by Moore[3], who was also responsible for the first solid evidence of the singularity[4]. It is uncertain whether special classes of problems (such as the Kaden formulation of the roll-up of a semi-infinite vortex sheet into a spiral[5]), which can be reduced to ordinary singular integro-differential equations in a similarity variable, have solutions. An asymptotic analysis of the spiral center for the Kaden problem has been given by Moore[6], and numerical results have been presented by Pullin[7], but attempts to improve the numerics or confirm their accuracy have so far failed[8].

Since vortex sheets are used to describe the separated flow past bluff and lifting bodies, the ill-posedness has serious consequences, as it implies that all numerical attempts to calculate sheet motion for more than a finite time will give results that are method-dependent, and criteria for "good" methods are subjective.

The singularity can be removed by replacing the sheet by a vortex layer of small but finite thickness δ, say, containing large but finite vorticity. It has been proved mathematically, and is physically obvious, that the solution is unique and exists for all time. (Global existence for the corresponding problem in three dimensions, where vorticity can be amplified by the stretching of vortex lines, is another matter.) Presumably, for $t < t_c$, the position of the layer will tend to the sheet whose evolution is described by (6). But what happens for $t > t_c$? A numerical study of the evolution of a vortex layer has been carried out by Pozrikidis and Higdon[9]. The results suggest as a definite possibility that the limit of the layer is a fractal; that is, the limit is a curve dense in a finite area with "dimension" greater than one. The smoothing of the fractal (by viscosity, in practice) will produce a vortex patch. There is, therefore, some mathematical justification for considering vortex patches produced by roll-up of vortex sheets, even though the process by which the patches are formed is not understood, and calculations of their initial structure are sensitive to the method used.

Patches. A vortex patch in two dimensions is a connected finite area (not necessarily but usually simply connected) containing vorticity. It is completely surrounded by irrotational fluid. Further, the vorticity in the patch is usually taken to be constant. Physically, this is probably a bad approximation, being based on experiment and on sheet roll-up calculations, but it is almost essential for analysis and numerical study of patch dynamics. How physically realistic the results of constant-vorticity patch calculations are is an open question.

The first non-trivial patch calculation was the 1876 solution by Kirchhoff for the steadily rotating ellipse. This solution was generalized by Moore and Saffman[10] for a steady elliptical patch (containing constant vorticity ω_0) in a uniform strain. Their results demonstrated the important concept of vortex *fission* by showing that steady solutions could not exist if the rate of strain ε exceeded $0.15\,\omega_0$. A numerical study (in which vortex patches were replaced by a large number of point vortices) indicated that a patch would be torn apart if placed in too large an irrotational straining field.

One of Lord Kelvin's motivations for studying vortex dynamics was his vortex theory of matter. It is curious that *fusion* also occurs for vortex patches. This behavior was demonstrated by Roberts and Christiansen[11], who showed that two equal circular vortex patches would amalgamate if placed too close together. Again, the method was that of replacing the patches by a cloud of point vortices, except that, instead of summing (5) to find the velocity field, a fast Poisson solver was employed to calculate the stream function. Perhaps the vortex theory of matter should not have been discarded quite so early!

A major advance in the analysis of patches was made by Deem and Zabusky's application of the water-bag method of plasma physics[12]. They showed that the velocity induced by a patch could be written

$$u + iv = \frac{\Gamma}{2\pi} \oint \log |(z - z')| \, dz' \tag{7}$$

as an integral over the boundary of the patch. Hence a partial singular integro-differential equation is obtained for the shape of the patches. The kernel in (7) is less singular than the kernel in (5), and there is no question here of ill-posedness. Deem and Zabusky successfully exploited this approach to generalize Kirchhoff's ellipses to find patches of triangular, square, etc., symmetry, and to calculate the shapes of co-rotating and counter-rotating vortex pairs. Other applications to vortex arrays were made by the present author and colleagues. Recently, the method of Schwarz functions (suggested by J. Jimenez) has been employed successfully[13].

A number of interesting results have been found in these calculations, and applications have been made to vortex fission and fusion and to the study of the properties of an infinite linear array of patches that models the turbulent mixing layer and a double staggered array that models the wake. More remains to be done; there are some open questions concerning bifurcation and new solutions, about which there is some mild controversy; but the problems here are primarily problems of detail and are resolvable. The main objection is to the assumption of constant vorticity in the patches.

Patches with sheets. A case of non-uniform vorticity is the uniform patch bounded by a vortex sheet. A special case, $\omega_0 = 0$, is sometimes called the stagnant or hollow vortex, since all the vorticity is concentrated in a sheet on the surface. The velocity of the fluid in steady motion is then constant on the boundary, from Bernoulli's equation (neglecting surface tension in the case of the hollow vortex), and the methods of free-streamline theory can be applied. Pocklington[14] found closed-form analytical solutions for a counter-rotating pair of stagnant vortices. Baker, Saffman and Sheffield[15] examined the structure and two-dimensional infinitesimal stability of a linear array of hollow vortices. All results are similar to those for uniform patches, except for details of certain limiting cases.

When $\omega_0 \neq 0$, the steady motions can be called Prandtl-Batchelor flows, after their theorem that the limiting process $t \to \infty$ followed by $\nu \to 0$ produces $\omega =$ constant in a region having closed streamlines. Existence now again becomes a difficult problem and is open in the case of flows with boundaries. In the absence of boundaries, a solution was presented by Sadovskii[16] for a touching vortex pair of counter-rotating vortices. Instead of the pair being roughly ellipsoidal, as is the case for touching patches without vortex sheets[17], there are now cusps at the ends. Prandtl-Batchelor solutions have been computed by Tanveer[18] when the vortices do not touch, but so far all attempts to confirm or repeat Sadovskii's calculation for the touching pair have failed. Accurate representation of the singularity at the cusps is the main stumbling block. This question is important in connection with the infinite-Reynolds-number limit of steady laminar flow past a bluff body, since Fornberg[19] in 1982 suggested, on the basis of his numerical solutions of the Navier-Stokes equations, that this limit might be a huge wake in the shape of a Sadovskii vortex.

III. Stability of steady vortices

One of the major advances in the current cycle of activity has been the ability to study the stability of vortices, using the new power to solve complicated equations or perform many iterations provided by large computers. This power, not available to scientists of earlier generations, can be and is misused to crunch numbers with no clear purpose, and usually with no increase in knowledge; but it has the potential to solve old problems and to open up new fields by discovering new phenomena.

Given a steady configuration of vortices, there are essentially three approaches to stability:

1. Linear theory can be used to find the spectrum of eigenvalues of infinitesimal disturbances. This procedure has the advantage that it can be done for both two- and three-dimensional disturbances, and it leads to linear eigenvalue equations that can usually be solved, with enough effort. A disadvantage is that it is limited to small disturbances and gives no information about the evolution of non-linear disturbances or about possible final states. Also, in inviscid systems, linear instability means oscillations, and weakly non-linear disturbances can still be unstable.

2. Initial-value calculations can be carried out numerically for a specified initial disturbance, and growth, decay, or boundedness can be observed. The advantage of this approach is that it deals with non-linear disturbances and gives information about long-term behavior. The disadvantages are, firstly, that the wrong kind of initial disturbance might be selected, and more unstable ones overlooked; secondly, that computations in three dimensions are not really possible by well-tested methods with available computing resources; and thirdly, that non-linear numerical instabilities may give spurious results.

3. Global analytical methods can attempt to establish stability to arbitrary disturbances by use of variational principles and related analytical ideas. This method originated in the work of Lord Kelvin. It was applied in a non-rigorous manner to the two-dimensional stability of arrays by Saffman and Szeto[20] and to the vortex pair by Saffman and Szeto[21]. These latter results have been criticized by Dritschel[22], who cast doubt on the validity of energy methods. These criticisms seem inappropriate. The problem in claiming global stability is the need to know the energies of all the steady states, and the possible error in Saffman and Szeto's work seems to be due to their incorrect calculation of the energies of steady configurations, and to their omission of some non-symmetric solutions, rather than to the concept itself. Rigorous applications by Wan and Pulvirenti[23] have shown the L^1 stability of the circular vortex patch to two-dimensional disturbances. Since the patch is of constant area, and its radius of gyration is conserved as a consequence of conservation of angular momentum, this result is no more than an application of Schwarz's inequality. In any event, L^∞ stability is of more physical interest, and is an open question.

There are now a number of interesting initial-value calculations (method 2) that have been done for a variety of different configurations. The 1972 pair calculation by Roberts and Christiansen[11] has been followed by many studies of pairs and arrays. The situation with regard to three-dimensional calculations is not so satisfactory. These are almost entirely Biot-Savart-law calculations for vortex filaments, neglecting core deformation and possible change of core structure (vortex breakdown). The first careful calculation of this type was done by Moore[24] for the long-time evolution of the instability of trailing vortices. The most interesting phenomena occur, however, when the vortex filaments get close together or become highly curved, and this situation is not describable by the Biot-Savart equation. For example, the fascinating joining and breaking of vortex rings described by Oshima and Asaka[25] is not describable by the thin-filament approximation. The origin of superfluid turbulence may be one field, however, which is describable in terms of the dynamics of vortex filaments[26].

Spectral calculations (method 1) have now been carried out for a number of configurations. For two-dimensional disturbances, there is a curious result for the stability of the finite-area Karman vortex street. In a frame moving with the street, relative to which the undisturbed vortices are stationary, the stream function for the steady state plus an infinitesimal normal-mode disturbance can be written

$$\psi(x, y, t) = \Psi(x, y) + e^{\sigma t} e^{2\pi i p x / L} \phi(x, y), \qquad (8)$$

where L is the streamwise distance between vortices, p is the subharmonic wave number of the disturbance, $\phi(x, y) \equiv \phi(x + L, y)$ is the spatially periodic eigenfunction, and σ is the unknown eigenvalue. The equations of motion imply an eigenvalue for σ of the form

$$\sigma = (\Gamma/L^2) \, \Sigma(p, h/L, A/L^2), \qquad (9)$$

where Σ is a dimensionless function of its arguments, h is the distance between the rows, A is the area of each vortex, and $\pm\Gamma$ is its strength. We denote the aspect ratio h/L by κ and the relative area A/L^2 by α. The circulations of the vortices in the rows must be equal and opposite, but the areas can be unequal. It is assumed that the rows are staggered, with each vortex opposite a point midway between vortices in the other row. The case $\alpha = 0$ was analyzed by Karman, who showed that $p = 1/2$ gave the most unstable disturbances and that $\mathrm{Re}\Sigma \neq 0$ (i.e., there is instability) for all κ except $\kappa_c = \cosh^{-1} 2 = 0.281$. There remains the question of the effect of finite α on this result. Saffman and Schatzman[27] argued erroneously that $p = 1/2$ was always the most unstable disturbance. For this case they showed that, for $\alpha > 0$, there exists a finite range of κ for which $\mathrm{Re}\Sigma = 0$, i.e., there is linear stability. These results are in agreement with initial-value calculations (method 2) by Christiansen and Zabusky[28]. However, Kida[29] showed by a perturbation analysis that the most unstable disturbance has $p \neq 1/2$ when $\alpha > 0$, but he still found a stabilizing effect. (Kida's method appears to be equivalent to the elliptical-vortex approximation[30] combined with the Betz result[31] that the centroid of a vortex patch moves with the average of the velocity over the patch.) A detailed recalculation of the eigenvalues by Meiron, Saffman and Schatzman[13] for arbitrary p and arbitrary α led to the remarkable conclusion that Karman's result for $\alpha = 0$ is qualitatively true for $\alpha > 0$, except that $\kappa_c = \kappa_c(\alpha)$, and the p of the most unstable disturbance varies with α for $\kappa \neq \kappa_c$. Remarkably, this result remains true when the rows are of unequal area[32]. A general result of this kind should have a simple analytical proof that does not rely on extensive calculations and computations, but none has yet been found. It is unlikely that these results could have been obtained using method 2 only. On the other hand, method 1 leaves nonlinear stability as an open question for finite area.

There have been some three-dimensional spectral calculations, but results are not yet extensive. The most significant are the calculations by Pierrehumbert and Widnall[33] on a mixing layer with a smooth but spatially periodic distribution of vorticity, and by Robinson and Saffman[34], who examined stability of arrays using the Biot-Savart approximation. The three-dimensional stability of a strained vortex was studied by Robinson and Saffman[35].

IV. Three-dimensional effects

As mentioned at the beginning of section II, attention has been concentrated on the case of flow in two dimensions because it is simpler mathematically, and because there appear to be real flows that can be modeled realistically with two-dimensional vortices. Recent work is beginning to cast doubt on this assumption by indicating that three-dimensional effects may in fact always be crucial.

Consider the Karman vortex street. Recent experiments by Couder, Basdevant and Thome[36] in soap films produce what is probably a genuinely two-dimensional wake. The observations show pairing instabilities and the formation of vortex couples (a pair of counter-rotating vortices) in laminar wakes at a Reynolds number of 950. This phenomenon is not seen in wind or water tunnels. Recent wind-tunnel smoke pictures by Cimbala[37], whose work was done in collaboration with H. Nagib and A. Roshko, show turbulent wakes at comparable Reynolds numbers. What is more interesting, at lower Reynolds numbers, where flow visualization has always shown a long street, Cimbala finds that the street decays quite quickly. The far-wake phenomena, such as the wake pairing reported by Taneda[38], appear to be due to instability of the parallel wake that exists after the initial street has decayed. The implication of these two experiments is that real wakes are controlled by three-dimensional effects, and that two-dimensionality occurs only when the viscosity is large enough to damp three-dimensional instabilities.

A related matter is the fact that a turbulent vortex is expected to produce negative mean vorticity[39,40]. Thus the use of uniform vortex patches to model turbulent vorticity distributions, as is done for turbulent mixing layers or trailing vortices, may be particularly bad; the behavior of vorticity distributions containing vorticity of both signs may be qualitatively different from the behavior of flows where vorticity of only one sign is present. A start has been made on establishing the fundamental properties of three-dimensional vorticity distributions, but it seems that work on this subject will be a major task for the 21st century.

Acknowledgment

I wish to thank the Office of Naval Research (Mechanics Branch) for their continuing support of the work on vortex dynamics in Applied Mathematics at the California Institute of Technology. Support is also being provided by the Department of Energy (Office of Basic Energy Sciences). Earlier work was sponsored by the Air Force Office of Scientific Research and the Army Research Office.

Appendix: vortex methods

It is perhaps appropriate to say a few words about the so-called vortex method, as there may be misconceptions about its relation to vortex dynamics. The vortex method constitutes a grid-free numerical approximation for the Euler or Navier-Stokes equations, with the vorticity field replaced by a linear sum of vorticity patches whose centroids move according to given equations. The patches are of constant shape (or are evolving in a predetermined manner) and are in general overlapping. See, for example, the application by Kuwahara and Takami[41] to patches and sheets. The method has attracted attention because there are convergence proofs that show, in appropriate limits and under sufficiently restrictive conditions, that the numerical approximation tends to the exact solution[42]. However, there has not yet been any convincing demonstration that the method is better than alternative numerical representations, or is a practical improvement for plane flows over the method of point vortices. The latter vortices have the property that they are a weak solution of the Euler equations, so even if they are not a good approximation to a continuous flow field, they have some interest in their own right. The available evidence suggests that the vortex method is rather inaccurate in practice. This is particularly so for the random-vortex method used to model the effects of viscosity at large Reynolds numbers, a method suggested in 1969 by D.W. Moore but not published because tests at that time showed large errors. Later work confirmed that the relative error in the calculation of the viscous diffusion is large unless the number of patches is large compared with the Reynolds number.

Vortex methods can be applied, of course, to the numerical calculation of initial-value problems in vortex dynamics, but to call an invariant patch that overlaps its neighbors (and does not deform) a vortex may be inappropriate and misleading.

References

1. H.J. Lugt, *Vortex flow in nature and technology* (Wiley-Interscience, 1983).

2. C.C. Lin, "On the motion of vortices in two dimensions," Univ. Toronto Appl. Math. Ser. No. 5, Univ. Toronto Press, 1943.

3. D.W. Moore, "Numerical and analytical aspects of Helmholtz instability," in Proc. XVI IUTAM Symp., Lyngby, 1984, 263-274 (1985).

4. D.W. Moore, "The spontaneous appearance of a singularity in the shape of an evolving vortex sheet," Proc. R. Soc. London **A365**, 105-119 (1979).

5. H. Kaden, "Aufwicklung einer unstabilen Unstetigkeitsfläche," Ing.-Arch. **2**, 140-168 (1931).

6. D.W. Moore, "The rolling up of a semi-infinite vortex sheet," Proc. R. Soc. London **A345**, 417-430 (1975).

7. D.I. Pullin, "The large-scale structure of unsteady self-similar rolled-up vortex sheets," J. Fluid Mech. **88**, 401-430 (1978).

8. J.C. Schatzman (private communication).

9. C. Pozrikidis and J.J.L. Higdon, "Non-linear Kelvin-Helmholtz instability of a finite vortex layer," J. Fluid Mech. **157**, 225-263 (1985).

10. D.W. Moore and P.G. Saffman, "Structure of a line vortex in an imposed strain," in *Aircraft Wake Turbulence*, edited by E. Olsen, A. Goldburg, and M. Rogers (Plenum, 1971), pp. 339-354.

11. K.V. Roberts and J.P. Christiansen, "Topics in computational fluid mechanics," Comput. Phys. Comm. **3** (Suppl.), 14-32 (1972).

12. G.S. Deem and N.J. Zabusky, "Vortex waves; stationary V states, interactions, recurrence and breaking," Phys. Rev. Lett. **40**, 859-862 (1978).

13. D.I. Meiron, P.G. Saffman, and J.C. Schatzman, "The linear two-dimensional stability of inviscid vortex streets of finite-cored vortices," J. Fluid Mech. **147**, 187-212 (1984).

14. H.C. Pocklington, "The configuration of a pair of equal and opposite hollow straight vortices, of finite cross-section, moving steadily through fluid," Proc. Cambridge Philos. Soc. **8**, 178-187 (1895).

15. G.R. Baker, P.G. Saffman, and J.S. Sheffield, "Structure of a linear array of hollow vortices of finite cross section," J. Fluid Mech. **74**, 469-476 (1976).

16. V.S. Sadovskii, "Vortex regions in a potential stream with a jump of Bernoulli's constant at the boundary," Prikl. Mat. Mekh. **35**, 773-779 (1971).

17. P.G. Saffman and S. Tanveer, "The touching pair of equal and opposite uniform vortices," Phys. Fluids **25**, 1929-1930 (1982)

18. S. Tanveer, "Topics in 2-D separated vortex flows," Ph.D. thesis, California Institute of Technology, 1983.

19. B. Fornberg (private communication).

20. P.G. Saffman and R. Szeto, "Structure of a linear array of uniform vortices," Stud. Appl. Math. **65**, 223-248 (1981).

21. P.G. Saffman and R. Szeto, "Equilibrium shapes of a pair of uniform vortices," Phys. Fluids **23**, 2339-2342 (1980).

22. D. Dritschel, "The stability and energetics of co-rotating uniform vortices," J. Fluid Mech. **157**, 95-134 (1985).

23. Y.H. Wan and M. Pulvirenti, "Nonlinear stability of a circular vortex," Comm. Math. Phys. **99**, 435-450 (1985).

24. D.W. Moore, "Finite amplitude waves on aircraft trailing vortices," Aeron. Quart. **23**, 307-314 (1972).

25. Y. Oshima and S. Asaka, "Interaction of two vortex rings moving side by side," Nat. Sci. Rep. Ochanomizu Univ. **26**, 31-37 (1975).

26. K.W. Schwarz, "Generation of superfluid turbulence deduced from simple dynamical rules," Phys. Rev. Lett. **49**, 283-285 (1982).

27. P.G. Saffman and J.C. Schatzman, "Stability of a vortex street of finite vortices," J. Fluid Mech. **117**, 171-185 (1982).

28. J.P. Christiansen and N.J. Zabusky, "Instability, coalescence and fission of finite-area vortex structures," J. Fluid Mech. **61**, 219-243 (1973).

29. S. Kida, "Stabilizing effects of finite core on Karman vortex street," J. Fluid Mech. **122**, 487-504 (1982).

30. P.G. Saffman, "The approach of a vortex pair to a plane surface in inviscid fluid," J. Fluid Mech. **92**, 497-503 (1979).

31. A. Betz, "Verhalten von Wirbelsystemen," Z. angew. Math. Mech. **12**, 164-174 (1932).

32. J. Kamm, "Shape and stability of two-dimensional uniform-vorticity regions," Ph.D. thesis, California Institute of Technology, 1987.

33. R.T. Pierrehumbert and S.E. Widnall, "The two- and three-dimensional instabilities of a spatially periodic shear layer," J. Fluid Mech. **114**, 59-82 (1982).

34. A.C. Robinson and P.G. Saffman, "Three-dimensional stability of vortex arrays," J. Fluid Mech. **125**, 411-427 (1982).

35. A.C. Robinson and P.G. Saffman, "Three-dimensional stability of an elliptical vortex in a straining field," J. Fluid Mech. **142**, 451-466 (1984).

36. Y. Couder, C. Basdevant, and H. Thome, "Solitary vortex couples in two-dimensional wakes," C.R. Acad. Sci. **299**, 89-94 (1984).

37. J.M. Cimbala, "Large structure in the far wakes of two-dimensional bluff bodies," Ph.D. thesis, California Institute of Technology, 1984.

38. S. Taneda, "Downstream development of the wakes behind cylinders," J. Phys. Soc. Japan **14**, 843-848 (1959).

39. S.P. Govindaraju and P.G. Saffman, "Flow in a turbulent trailing vortex," Phys. Fluids **14**, 2074-2080 (1971).

40. P.G. Saffman, "The structure of turbulent line vortices," Phys. Fluids **16**, 1181-1188 (1973).

41. K. Kuwahara and H. Takami, "Numerical studies of two-dimensional vortex motion by a system of point vortices," J. Phys. Soc. Japan **34**, 247-253 (1973).

42. T.J. Beale and A. Majda, "Vortex methods for fluid flow in two or three dimensions," Contemp. Math. **28**, 221-229 (1984).

Dynamical System Theory and Simple Fluid Flow

Albert Libchaber
Professor of Physics
University of Chicago, Chicago, Illinois 60637

I have a close friend from École Normale who used to visit Caltech, and who told me of the magnetism of Liepmann and how attractive his scientific personality is. I have not known Professor Liepmann very long, but last year I met him in Japan and I got caught too.

What I want to talk about today is non-periodic flow described by deterministic equations. Yesterday, Dr. Narasimha said that I may shed some light on the problem of turbulence. I am not too optimistic. Physicists have often tried to understand turbulence in the past, from Heisenberg to Landau to Feynman, but physicists come and go and the problem of turbulence remains. What I hope to show in this historical sketch is that deterministic systems leading to chaotic behavior are interesting in themselves and perhaps are relevant for the coherent structures that are observed in many flows.

The subject itself is not a new one. Among partial differential equations, the famous Burgers equation has interesting chaotic solutions. In Sommerfeld's textbook[1] there is a beautiful analysis of the problem, with Sommerfeld himself asking questions about the relation of these solutions to turbulence. As far as I am concerned, the subject first attracted my interest with the conjecture by Ruelle and Takens in 1971 about strange attractors. Ruelle has written a clear paper[2] on the subject. What it is about is the following: In the Landau picture of turbulence, as the control parameter is increased, the fluid is stressed more and more. More and more limit cycles or oscillators are destabilized, and eventually the system becomes chaotic. The conjecture by Ruelle and Takens was that if we have a three-torus, which means a system where we have destabilized three oscillators, we could be close to what they called a strange attractor. This is true only for dissipative dynamical systems that contract in phase space. A strange attractor is an attractor that is very sensitive to initial conditions. If we have two very close nearby trajectories, as close as we want, they will diverge exponentially with time. This sensitivity to initial conditions leads to chaotic behavior in the solution of the problem. The conjecture by Ruelle and Takens triggered a lot of interest in the physics community. Then Gollub and Swinney, in the United States, found in a Couette flow experiment[3] that, after two oscillators were excited, as they increased the control parameter they entered into a chaotic regime, defined by the appearance of broad-band noise in the experimental Fourier spectrum.

That was our knowledge to begin with. We also knew about the beautiful work done in 1963 by the meteorologist Lorenz[4]. In a paper which is about deterministic non-periodic flow, he presented and explained in a clear way the strange-attractor idea. What Lorenz did was to try to create a model of the atmosphere. He simplified the equations of the Rayleigh-Benard problem by truncating them to obtain a set of three ordinary differential equations with three parameters; the Prandtl number of the fluid, a control parameter, and a wave number of the convective roll structure. Working with these three coupled equations, Lorenz showed that indeed one can have a chaotic solution and that strange-attractor-like behavior is present. Consider a Benard experiment, with a fluid layer heated from below, and suppose that the heat diffusivity is not too large. Physically, what Lorenz was saying was that hot spots may develop in the fluid at a given value of the control parameter. These hot spots are advected by the motion, and, if the convection is fast enough, they have no time to cool down near the top plate before they descend again. But a hot spot moving downward is unstable, and the result is the Lorenz chaotic attractor. We thus have a physical model which shows that something is relevant in the Ruelle-Takens conjecture. If we start in a very simple way in a controlled experiment, by changing the constraints, we can destabilize limit cycles, and we can reach a chaotic state with a small number of limit cycles.

A further important step was the Feigenbaum analysis of period doubling in terms of mapping theory. This period-doubling model is not a new one. In France, at the beginning of the century, Fatou and Julia did a lot of work on non-linear mapping. During the war, for example, the same mapping considered by Feigenbaum was used by von Neumann and Ulam[5] to serve as a random-number generator for early computers. The important discovery by Feigenbaum[6] was that the transition to a chaotic state is equivalent to a second-order phase transition. Wilson entitled his article[7] in Scientific American about second-order phase transitions, "Problems in physics with many scales of length." In the transition to a chaotic state through period doubling we have instead a problem with multiple scales of time.

How is this relevant to experiments? Suppose we have an experiment with one bifurcation, say a Hopf bifurcation to a time-dependent state. What is the relation to the evolution of fixed points? We know the importance of the limit cycle from the early work of Poincare[8]. If we read the "Mécanique Celeste," we find that everything I have to say is more or less in germ in that book. If we have a limit cycle and make a cross-section, what is important is that the intersection point is a fixed point. Each time the cycle comes back, it comes back to the same point. When we go from one point to two points, the trajectory takes twice as much time on the average to come back. This is the beginning of period doubling.

All this is very nice, pure mathematics. From Poincaré to Arnold there is an enormous amount of mathematical work on this problem. Is this relevant to experiments? One of the simplest possible experiments[9] is the Rayleigh-Benard experiment. I feel close to it because in 1900 Benard was in an institute nearby the École Normale. In the Benard experiment, we heat a fluid from below, and we get a first bifurcation, from heat diffusion to heat convection. As we keep increasing the control parameter (the temperature difference) we find other bifurcations, whose theory has been developed in a remarkably complete form for all types of fluids by Busse[10].

Then we had a surprise. What turns out to be very relevant in the problem is: What is the size of the box? How many cells do we have in our experiment, a small number or a large number of cells? In the Fourier spectrum shown in Fig. 1, for example, we have on the y-axis the logarithm of a temperature signal or a velocity signal. If we have a small number of cells, we have a very sharp, well-defined peak, with one or two oscillators, quasi-periodic or locked, and essentially no noise. But if we move to a large-aspect-ratio cell, we find that as we increase the control parameter what we have is still a well-defined oscillator, which in this case arises from the oscillatory instability predicted by Busse, but then a lot of noise comes into the problem. It becomes a problem with a large number of degrees of freedom, and this is accomplished by changing the aspect ratio, the size of the cell.

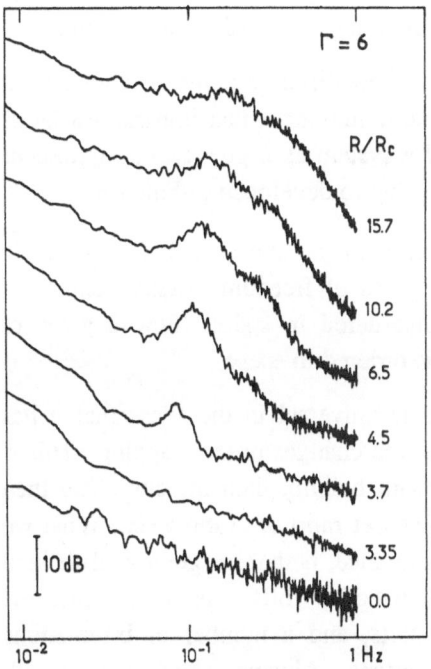

FIG. 1. Large-aspect-ratio convection in liquid helium. Power spectra (log scales) for various Rayleigh numbers. The curves have been translated along the vertical axis for clarity. The 0.0 curve shows the $1/f$ noise of the bolometer. The broad peak is associated with the oscillatory instability.

Now I come back to Benard's historical experiments with his collaborator. In 1913 Dauzère[11] published pictures of Benard cells. The pictures have nothing to do with an ordered convective state. They are full of curvature and dislocations; in fact, they might be the first observation of dislocations in science. It is these disordered patterns that lead to noisy behavior at onset. Now we know, from experiments done by a number of physicists, that a real picture in a large-aspect-ratio convective cell would look very disordered[12]. We have a lot of defects: dislocations, grain boundaries, and various structures. We can try to simplify them and work with one dislocation. But the problem remains complex. There is a disordered pattern in space. What we wanted was an ordered pattern in space. Again I go back to Benard, who published in the 1930's a paper in which he wanted to show the relevance of convection to meteorology, so he worked in air. He was able to realize large ordered patterns even in a large-aspect-ratio cell. In order to do this, Benard added a very small flow field in the horizontal direction. By doing tricks like this, or by having a small aspect ratio, we can align the pattern, and with an aligned pattern we can play our game of a small number of dimensions. Now Benard, as he increased the control parameter, found that his structure changed to an oscillatory pattern. This was the first encounter with the oscillatory instability that has since been analyzed by Busse. Essentially, what we want to do is start with a small number of rolls and look at the dynamics. To test dynamical system theory we need a small-aspect-ratio cell with perhaps six cells as a maximum.

Let me say immediately that the evolution can follow different paths. Figure 2 is an amusing picture drawn by Paul Martin at one of our conferences that illustrates a lot of what I will say. It shows the general evolution of a system as it goes from a quiescent state, say $Ra = 0$, to a very large Rayleigh number, up to developed turbulence.

In our game we talk about Poincaré maps, return maps, and so on. But the main problem is many degrees of freedom and few degrees of freedom. Many degrees of freedom means that we have structure that is disordered in space. Few degrees of freedom means that we start from a structure that is ordered in space.

Remember the period-doubling regime, which is universal in the sense that it has specific scaling properties and also shows robustness to changes in the mapping. This is the first thing we studied. We did experiments with helium, then mercury, and then potassium. In all cases we used small cells, with aspect ratio of at most six. What we see in real time, as we increase the temperature difference, is that we get a well-defined oscillation. This is a transverse motion of the convective rolls. As we increase the temperature difference further, period doubling starts, and a number of bifurcations occur. We can see up to 4 period-doubling bifurcations. Always, when we play with the Fourier spectrum on a logarithmic scale, we can blow up everything. We start from a situation where we have essentially no noise, say 70 db signal-to-noise ratio. We increase the control parameter and see the first period doubling. We increase the control parameter further and find the next bifurcation. We can see up to four bifurcations, as shown in Fig. 3.

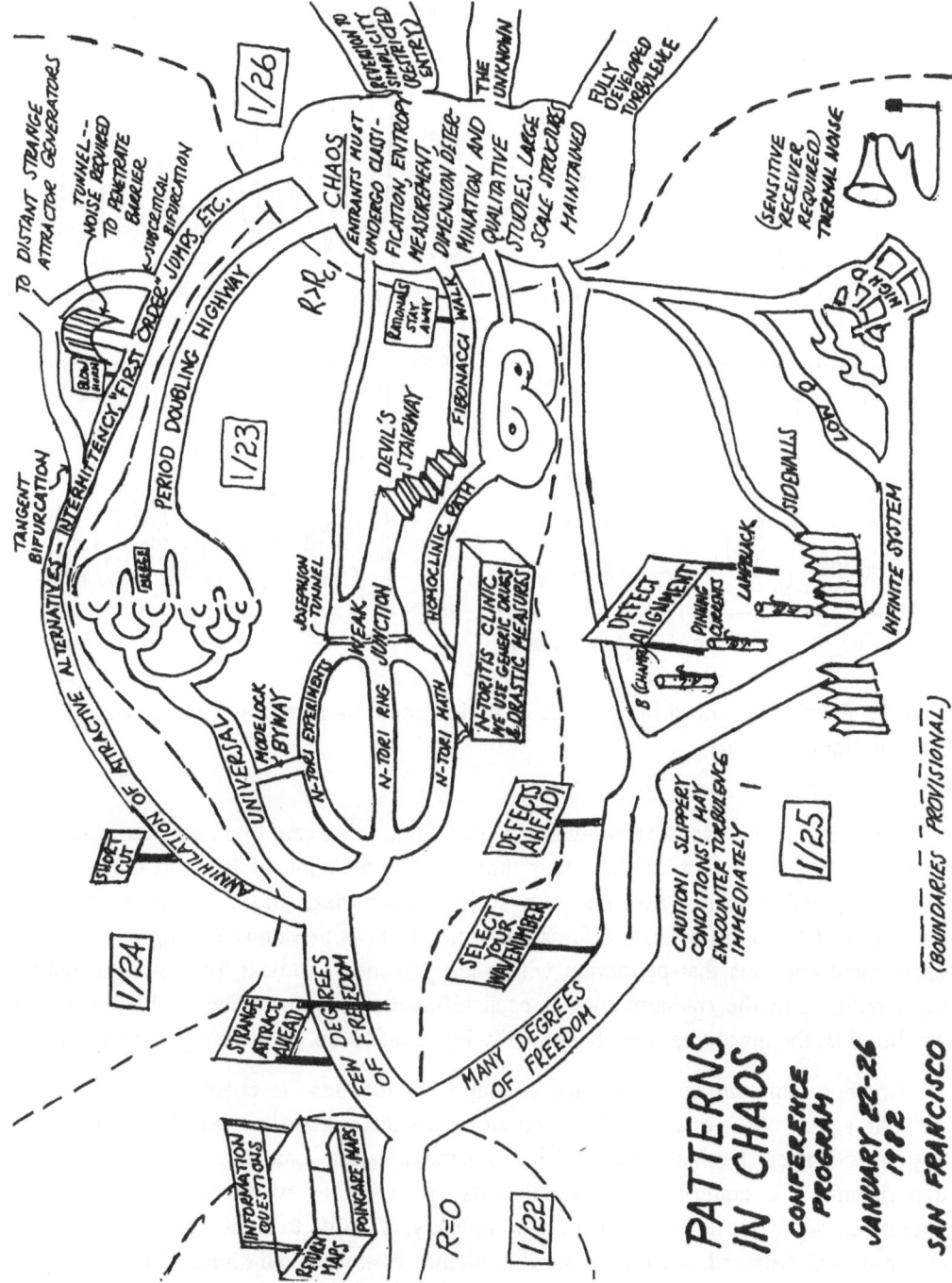

FIG. 2. Patterns in chaos, an amusing view of various possible routes (courtesy of Paul Martin).

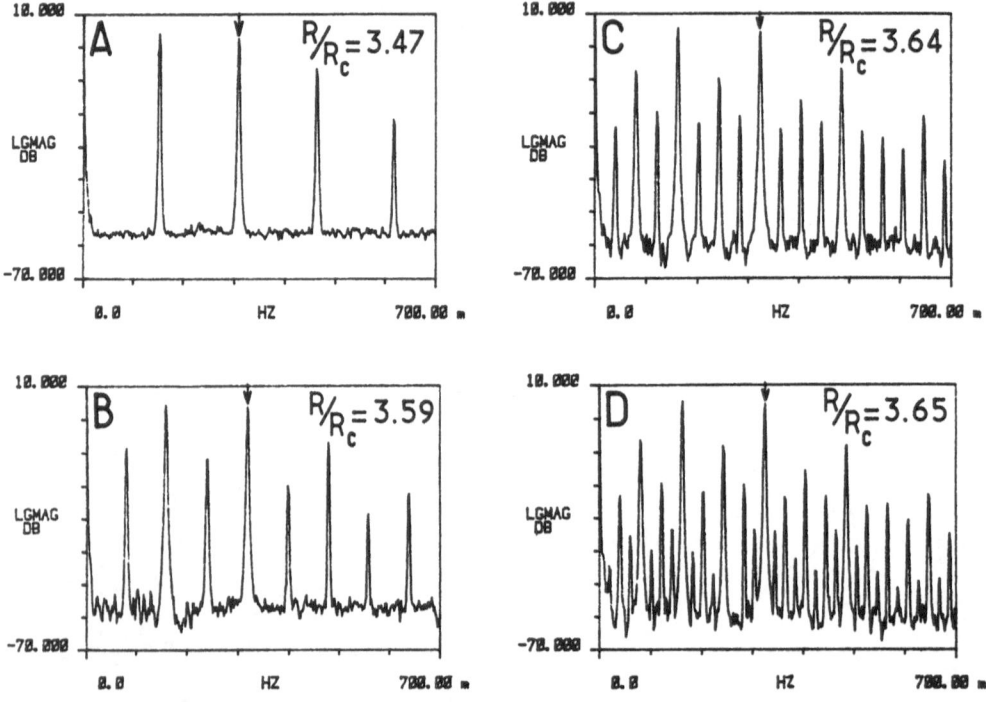

FIG. 3. Fourier spectra of the cascade of period doubling for a Rayleigh-Benard experiment in liquid mercury.

Essentially, what we observe, when we carry out a Poincaré cross-section, is that we have a fixed point which doubles four times. The prediction of Feigenbaum was that this is a well-defined transition with scaling properties, and there is a universal parameter at the onset of each bifurcation which follows a geometric progression. Our experiment confirms that prediction within ten percent. The next thing was to look at the amplitude of the subharmonics at each bifurcation. Each period-doubled signal is smaller than the preceding one, theoretically by about 13 db. We were getting 14 db.

Another thing the model shows is that once the flow is chaotic it can become laminar again. As we increase our control parameter, there are windows where the system becomes laminar again. This relaminarization can also be seen in our experiment. We could see various periods appearing as we increased our control parameter, while the noise disappeared completely. Finally, to show that the small fluid cell was well behaved, we know that this picture is always self-similar. For example, when we are in the laminar region, we get out of the laminar region by period doubling again. The same picture reproduces itself for any window. That could be seen, for example, in the region where period three appeared in the experiment.

Clearly the problem in terms of the dynamical-system approach is more complicated. When we have a limit cycle and we change the control parameter, the system can destabilize by period doubling. But, in fact, if we look at Fig. 2 we see that when we have a few degrees of freedom, we have three main roads. These three main roads are very well understood mathematically. They correspond to three types of bifurcations that can happen to a limit cycle. If we have a limit cycle, which is an oscillation in time, and look at a Poincaré cross section, we find a fixed point. The first possibility is that this fixed point becomes two points. The second possibility is that the fixed point disappears, which means that we lose our limit cycle, and we lose our oscillation. It disappears by interaction with an unstable fixed point. The result is what is often observed in fluid mechanics; intermittency. As we come close to the disappearance, when those two points interact, we find that there are bursts of noise. But before I show that, let me say what is the third possibility, once we have a limit cycle. When we increase the control parameter, this fixed point can bifurcate through Hopf bifurcation to a limit cycle. In that case, we are in a quasi-periodic state. This is the beginning of the old Landau picture of first oscillator, second oscillator. These are the only three possibilities.

These three possibilities are shown in Fig. 2. There is universal period doubling. There are n-torii states, which means quasi-periodicity. There is annihilation of the attractor, because the limit cycle loses stability. This intermittency in the case of dynamical systems was explained by Pomeau and Manneville[13], and there is an elementary case where we can understand the process. If we again think of a mapping, we can have a situation where as we iterate we have a fixed point which is stable, and another fixed point which is unstable. As we iterate, we move away from one point and toward the other. When we change our control parameter, we move the curve with respect to the bisectrix. Finally, we can reach a state where the two points have coalesced. When we are in this situation, as we do our iteration we stay a very long time near what was a fixed point. Then we go somewhere else in a complex trajectory, and when we come back we stay again a long time near this fixed point. We will have intermittency in the sense that the laminar regime will be present, but as we increase our control parameter and move away we find less and less laminar period. For example, we observed this phenomenon in liquid helium, as shown in Fig. 4. Intermittency also has scaling properties, which were developed by Pomeau and Manneville and have now been found in a large number of systems. Remarkable work was also done by Bergé and his group on the question of intermittency. Outside of period doubling and intermittency, a large class of experiments will fall in a regime of quasi-periodicity.

Once we are in a quasi-periodic state, the regimes we can get are extremely numerous. But, in general, the result of all the observations so far is that when we do an experiment of this sort, as we increase our control parameter it is very hard to go beyond three or four limit cycles and we always fall in a chaotic state.

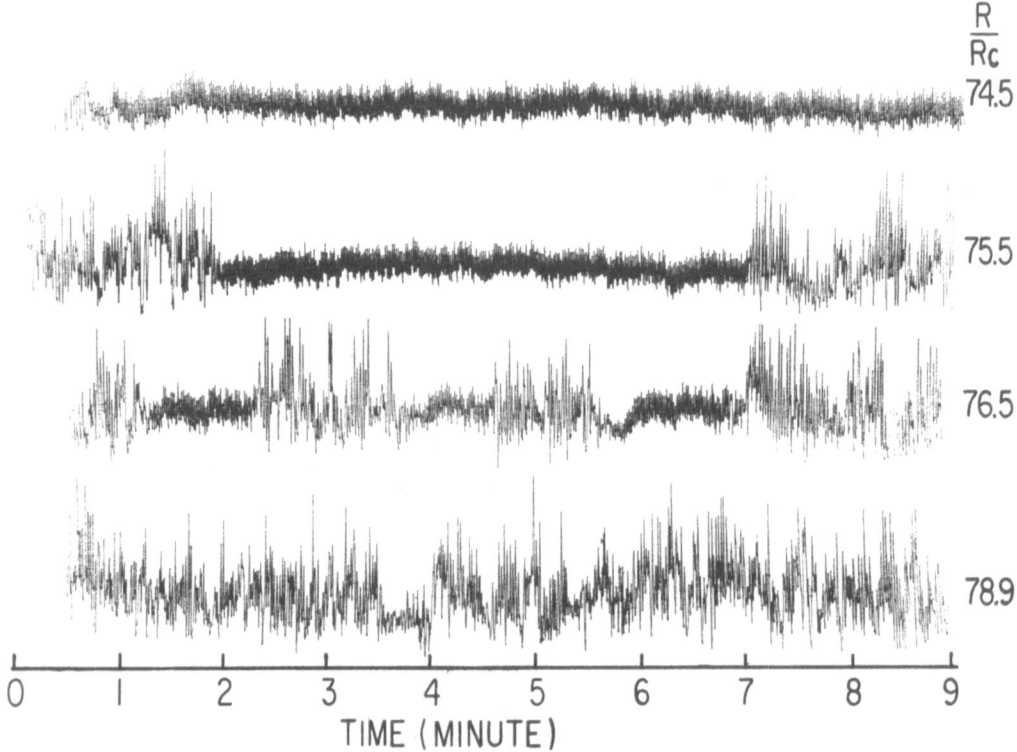

FIG. 4. Intermittency in liquid helium. Evolution of the time signal near the chaotic transition. The motion is quasi-periodic between the noisy bursts.

There are very many interesting problems in this quasi-periodic regime. Mathematically, one of the most interesting is what happens with two incommensurate frequencies when we keep the ratio of the two frequencies constant and increase our control parameter. We have a system with two incommensurate frequencies. They are not locked. We try to keep this ratio fixed and we change the control parameter. There is again a transition with scaling exponents. The theory of this transition was done by two groups[14]. I will not go into it, but will say that recently we have observed this transition and measured the exponents[15].

In these problems with a few degrees of freedom, we can see that there are three elementary bifurcations, and from there we can reach a chaotic state. In this chaotic state, following again the Poincaré cross-section, we can look at the strange attractor and the dimensional fit, and obtain a fractal dimension. This is being done currently in different laboratories. The whole concept seems to be quite relevant.

Finally, I would like to say something about the problem of turbulence. The game I have played here is just simple dynamical-system mathematics. In a situation where we start from an ordered state in space, this dynamical-system approach is valid. It could very well be that in some problems in turbulence; for example, in the origin of coherent structure, we will find some of this behavior. I said at the beginning that when we have disorder in space, we have a difficult problem. We already have noise at onset, with a large number of degrees of freedom. What we would like to understand now is the problem of many degrees of freedom.

A lot of research work now is applied to understanding phase instabilities. Suppose we have a number of clocks that are weakly coupled in phase, and we want to understand the evolution of the one-dimensional problem in space. There is a book by Kuramoto that describes such problems. Many physicists are working now with the Kuramoto-Sivashinsky equation, which deals with very-large-aspect-ratio one-dimensional problems. It is an equation that can describe the evolution of the phase of the structure.

Let me conclude by saying that in Lord Rayleigh's book[16] "Theory of Sound," there is a chapter on nonlinear differential equations where, for the first time, parametric amplification is discussed. Lord Rayleigh had everything in his hand. We can imagine what our understanding might be now if he could have pushed a little further at that time.

References

1. A. Sommerfeld, *Mechanics of Deformable Bodies*, Lectures in Theoretical Physics, Vol. II (Academic Press, 1950).

2. D. Ruelle, Commun. Math. Phys. **82**, 137 (1981).

3. J.P. Gollub and H.L. Swinney, Phys. Rev. Lett. **35**, 927 (1975).

4. E.N. Lorenz, J. Atm. Sci. **20**, 130 (1963).

5. S.M. Ulam and J. von Neumann, Bull. Am. Math. Soc. **53**, 1120 (1947).

6. M.J. Feigenbaum, Commun. Math. Phys. **77**, 65 (1980).

7. K.G. Wilson, Sci. Am. **241**, No. 2, 158 (1979).

8. H. Poincaré, *Les Méthodes Nouvelles de la Mécanique Céleste*, Vol. 3 (Dover, 1945).

9. A. Libchaber and J. Maurer, in *Nonlinear Phenomena at Phase Transitions and Instabilities*, edited by T. Riste (Plenum Press, 1982).

10. F.H. Busse, Rep. Prog. Phys. **41**, 1929 (1978).

11. M.C. Dauzère, C.R. Acad. Sci. (Paris) **156**, 1228 (1913).

12. V. Croquette, Thesis, Paris (1985).

13.   Y. Pomeau and P. Manneville, Commun. Math. Phys. **74**, 189 (1980).

14.   For example; M.J. Feigenbaum, L.P. Kadanoff, and S.J. Shenker, Physica **5D**, 370 (1982).

15.   J. Stavans, F. Heslot, and A. Libchaber, Phys. Rev. Lett. **55**, 596 (1985).

16.   Lord Rayleigh, *The Theory of Sound*, Vol. I, Chapt. 3 (Dover, 1945).

Turbulence Research by Numerical Simulation

A. Leonard†
NASA Ames Research Center, Moffett Field, California 94035
and
W. C. Reynolds
Stanford University, Stanford, California 94305

I. Introduction (WCR)

Our task is to describe recent progress and present prospects for research in turbulence using numerical methods. This is quite a challenge, because we know that Liepmann has not been a strong advocate of computational turbulence, or "compulence," as it has sometimes been called -- the incompressible version presumably being "incompulence." We want to recognize several objectives in exploring turbulence by numerical methods, and to order these in priority as follows:

1. To gain understanding and insight into the physics of turbulence, so as to complement the insight obtained from experiments and analysis; for example, by helping the experimentalist to understand what it is that has been observed experimentally, and by helping the analyst to explore solution space in more detail.

2. To provide special "data" for guiding, evaluating, and calibrating simpler predictive models of turbulent flows.

3. To predict turbulent flows by numerical simulation.

We believe that too much attention has been paid to the last of these objectives, whereas the real profit is more likely to come from work aimed at the first two.

In the words of Don Coles, "A computation that predicts a flow is not an explanation. What is required is an explanation." We hope to show that explanations can come, have come, and will come from research on turbulence by numerical methods.

We will present a series of selected examples to illustrate our points. We will emphasize three aspects of this work:

1. How does one learn from numerical simulations?

---

† Present address: California Institute of Technology, Department of Aeronautics 301-46, Pasadena, California 91125

2. What can be learned from numerical simulations?

3. What might be learned from numerical simulations?

Tony will begin by discussing what has been learned by study of some special, simple flows that would be very difficult to set up in the laboratory.

II. Interacting vortex rings (AL)

It is well known that an isolated vortex ring with a thin core in an ideal fluid will propagate at constant speed. As discussed below, certain fat vortex rings will also move at constant speed. What happens when two vortex rings collide along a common axis? We can simulate this process to very high accuracy for a special class of vortex rings, those whose core is defined by a patch of vorticity within which the vorticity is proportional to the distance r from the axis of symmetry. In this case, the vorticity transport equation,

$$\frac{D(\omega/r)}{Dt} = 0 \ , \tag{1}$$

is satisfied exactly for all interior points. All we need to do is to track the core boundaries, which are closed curves in the (r,x) plane. Thus we can reduce a two-dimensional problem to the problem of tracking the one-dimensional boundaries of the patches. By doing so, we obtain a considerable increase in resolution with the same computational resources. Starting with the Biot-Savart law, Shariff et al.[1] (and independently Pozrikidis[2]) have shown that the equations of motion for these bounding curves can be reduced to the form

$$\frac{\partial X_i(\xi,t)}{\partial t} = \int G_{ij}(r,r',x-x') \frac{\partial X_j(\xi',t)}{\partial \xi'} d\xi' \ , \tag{2}$$

where ξ and ξ' are Lagrangian coordinates and the kernel G is composed of complete elliptic integrals. Similar methods have been used in the planar case to investigate unsteady[3,4] and steady[5,6] flows.

Some time ago, Norbury[7] computed, by a different technique, a one-parameter family of vortex rings in steady translation with the property $\omega \sim r$. The core shapes of some members of the family are shown in Fig. 1. At one extreme of the family is Hill's spherical vortex, and at the other extreme are vortex rings with nearly circular cores of vanishing cross-sectional area.

Using marker points to track the boundaries, together with suitable interpolation and integration schemes, we can integrate Eq. (2) and determine what happens when two vortex rings collide. In a computation especially commissioned for this symposium, K. Shariff computed the evolution of two identical Norbury rings approaching each other along a common axis. From simple arguments based on the Biot-Savart law for thin

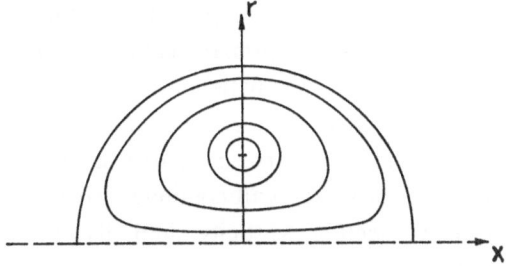

FIG. 1. Boundaries for some members of Norbury's family of steady vortex rings[7].

cores, we know that the rings will slow down and increase in radius, and that their cores will decrease in area as they approach each other. But what happens during and after the collision? How do the cores of the vortices deform? What is the final asymptotic state (if there is one) of the interaction? Figs. 2a-2g show some results of the simulation that bear on these questions. Before the interaction, the cores are slightly elongated in the direction of travel (Fig. 2a). During the interaction, the major axes tilt and eventually rotate 90 degrees from their original direction (Figs. 2c, 2d). The two cores

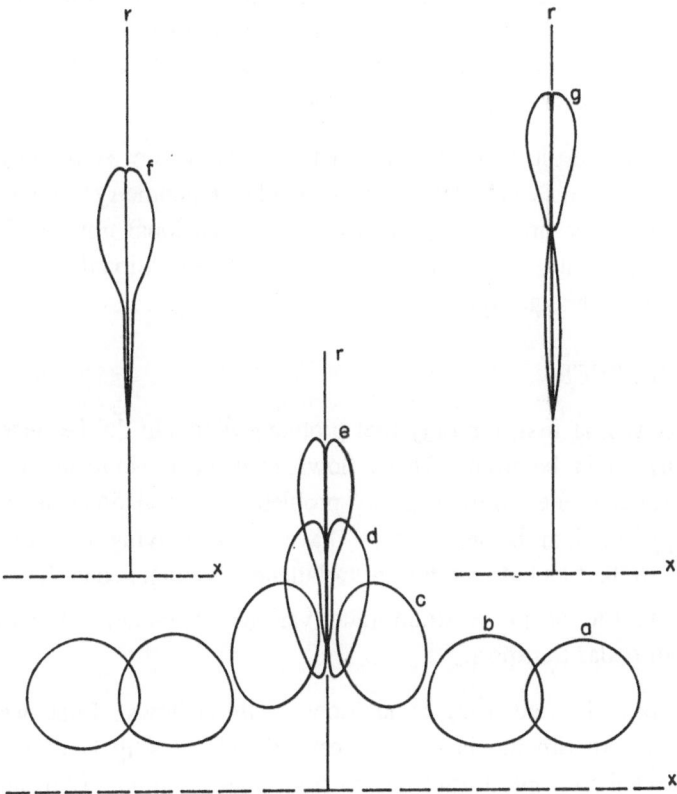

FIG. 2. Collision of two vortex rings (computation by K. Shariff).

then propagate roughly as a vortex pair in the radial direction, except that a significant spike of vorticity is left in the wake (Figs. 2e, 2f, 2g). The primary cores appear to pinch off the spike and leave it behind. Of course, because the spatial derivatives of the velocity are bounded, the spikes cannot really pinch off. The two bounding curves cannot actually touch, but they can become exponentially close. Analysis of the energetics of stretching indicates that the generation of spikes will continue. That is, a self-similar core shape cannot persist without ejecting vorticity into its wake. Experiments by Oshima[8] support these observations.

At the time of the symposium we thought that core oscillations observed during the collision were merely of intellectual interest, with no relevance to experiments. However, Shariff et al.[9] have recently argued that these oscillations play a major role in the acoustic signal produced by vortex-ring interactions. This is because the far-field acoustic pressure signal at low Mach number[10] is proportional to

$$\frac{d^3}{dt^3} \iint x\, r^2 \omega\, dx\, dr \quad . \tag{3}$$

The signal depends on oscillations involving the ellipticity of the core and its orientation, as well as the motion of the core centroid. Acoustic pressures computed using Eq. (3), together with the results of contour-dynamics calculations, show much better agreement with the colliding-ring experiments of Kambe and Minota[11] than pressures computed on the basis of a circular core of variable diameter.

What about the spikes that result from the interaction? They are commonly observed in unsteady contour-dynamics calculations both in planar geometry[4] and in axisymmetric geometry[1,2]. Are they simply an artifact of two-dimensional flows with vortex patches, or are they important small-scale features in three-dimensional flows? We do not yet know the answer to this question.

III. Homogeneous turbulence (WCR)

Homogeneous turbulence was, at first, the only real problem we could do, because periodic boundary conditions could be used. These flows have been simulated by several investigators. In particular, the isotropic decay problem has been done many times. The return-to-isotropy problem is one that we have been studying recently. Other problems that we have studied include rotation of turbulence, homogeneous shear, homogeneous strain, axisymmetric or plane strain (and various combinations), and compression, all possibly with scalar transport.

Let us review what has been learned using simulations of these flows. First, we consider rotation. Some nice experiments have been carried out by Wigeland and Nagib[12], in which they subjected isotropic turbulence to solid-body rotation. With no rotation, they obtained the middle curve in Fig. 3 for the decay of turbulence intensity.

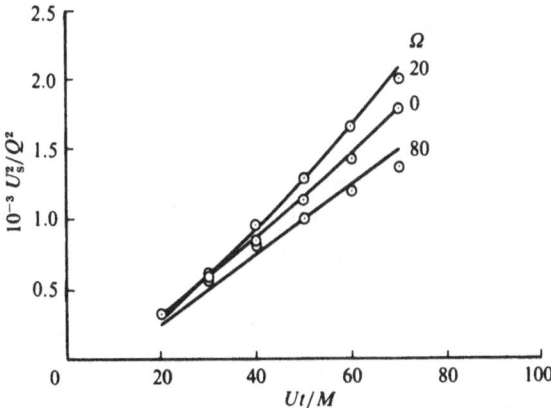

FIG. 3. Decay of rotating homogeneous turbulence according to experiment[12] (symbols) and numerical simulation[13] (solid lines). Initial dissipation rate varies from case to case (courtesy of Journal of Fluid Mechanics).

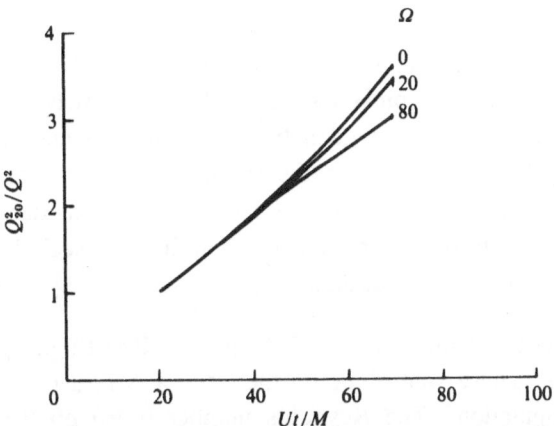

FIG. 4. Decay of rotating homogeneous turbulence by numerical simulation[13] (courtesy of Journal of Fluid Mechanics). Same initial conditions for all cases.

When they rotated the turbulence at 20 radians per second, they obtained the top curve. Then they rotated it faster and obtained the bottom curve. It was not clear from these experiments what really was happening. This example illustrates what often can be done with computations that can not be done with an experiment. In particular, it is easy in the computations to control the initial conditions. Shown in Fig. 4 is the result of a large-eddy simulation by Bardina[13] in which he starts all the computations with exactly the same initial conditions. Now we can see that the rotation reduces the rate at which the energy decays; i.e., rotation reduces the dissipation rate. The question is, how? What rotation does is apply a gyroscopic torque to the vortex filaments, making them line up with the main axis of rotation. This action increases the length scale along the axis of rotation, and the increased length scale slows the energy transfer to the smaller eddies. Consequently, the dissipation rate is reduced. This explanation was initially suggested by a detailed study of the simulations.

Now, let us discuss scalar transport[14]. In a homogeneous shear flow, we must have a velocity gradient; we can also have a temperature gradient. Stan Corrsin found experimentally that the $\overline{u'T'}$ correlation, which is the essentially the axial heat flux in this flow, is actually a little more than twice as large as the cross-stream heat flux, $\overline{v'T'}$. In this situation, we have a strong cross-gradient heat flux in the x-direction being driven by a temperature gradient in the y-direction. I know that gradient-transport theory is not loved here at Caltech. However, dimensional analysis is, and if we do a dimensional analysis for this problem we soon convince ourselves that the correlations have to scale on $\partial T/\partial x$. There is no other way that they can scale. What we can do with direct simulations is to study the anisotropic form of the diffusion-coefficient tensor. The gradient law suggested by dimensional analysis does not necessarily mean that the process is gradient diffusion. As we will see, the process is really vortex transport.

Joel Ferziger and his student C-T. Wu studied homogeneous compression[15], and their conclusions are short and simple. If we compress the fluid relatively slowly, so that the product of the strain rate times the energy divided by the dissipation rate is small, say one-tenth or smaller, then the time scale for the turbulence is k/ε, as turbulence modelers always assume. However, if we impose larger but still modest dimensionless deformation rates of order unity for this parameter, then the time scale is quite different from k/ε. Modelers should worry about that.

Next I will discuss simulations done by Moon Lee[16], with help from Bob Rogallo, of homogeneous turbulence with strain. These were done on the Cray X-MP, using a 128^3 mesh deforming during the computation. The Reynolds number based on the Taylor microscale is about 50, which according to Narasimha[17] is the beginning of real turbulence. The strain stretches the vortex filaments and lines them up with the direction of positive strain rate. We then shut off the strain rate and watch the vortex filaments try to return to a state of isotropy. Whatever measure of anisotropy is used, the vorticity anisotropy is very, very large. Of course, we should expect this. If we pull vortex filaments out by straining the flow, we essentially make the vorticity one-dimensional. Presumably, the vorticity would like to be three-dimensional. Strain causes a very rapid increase in the vorticity anisotropy and a corresponding increase in the anisotropy of the Reynolds stress. The lore of the subject tells us that when we remove the strain rate the small scales will relax toward isotropy very quickly while the large scales will relax toward isotropy at a much slower rate. The simulations show that the vorticity starts to relax very quickly, but after a while the vorticity anisotropy becomes locked to the Reynolds-stress anisotropy, and the two of them decay at the same rate. Thereafter, the small-scale anisotropy, as reflected in the vorticity, is in fact a little larger than the large-scale anisotropy, as reflected in the Reynolds stress. The small scales are decaying at the same rate as the large scales, not at a faster rate as turbulence lore would suggest. I think this is the most interesting result that has come out of these simulations. What it suggests is that small-scale structures are really a much more important dynamic part of large-scale structures than we may have thought. We will come back to this point later.

Now to shear flows. These have been studied by Rogers and Moin[18], again using the Rogallo code. A shear flow is simply a plane strain and a rotation, and we have seen that rotation tends to turn the vortex filaments and to align them with the axis of rotation. At the same time, the strain tends to pull them out along a 45-degree line. The question is, what does the structure of the vorticity field in a shear flow look like, particularly along a plane inclined at 45 degrees to the flow direction? A vorticity-vector plot (Fig. 5) shows zig-zag vortical structures that are really not very much different from the kind of vortex structures we see in the mixing layer. Rogers has made this observation quantitative by constructing a histogram of the vortex structures, with each sample weighted by the magnitude of the vorticity. We see in Fig. 6 that the predominant vorticity is aligned at 45 degrees or − 135 degrees to the flow direction. Thus vortex filaments in a shear flow, even a homogeneous shear flow, are strongly anisotropic and are aligned with the major principal axis of the strain tensor. This result suggests that large-scale structures in shear flows might be thought of as a lot of small-scale structures moving as a unit.

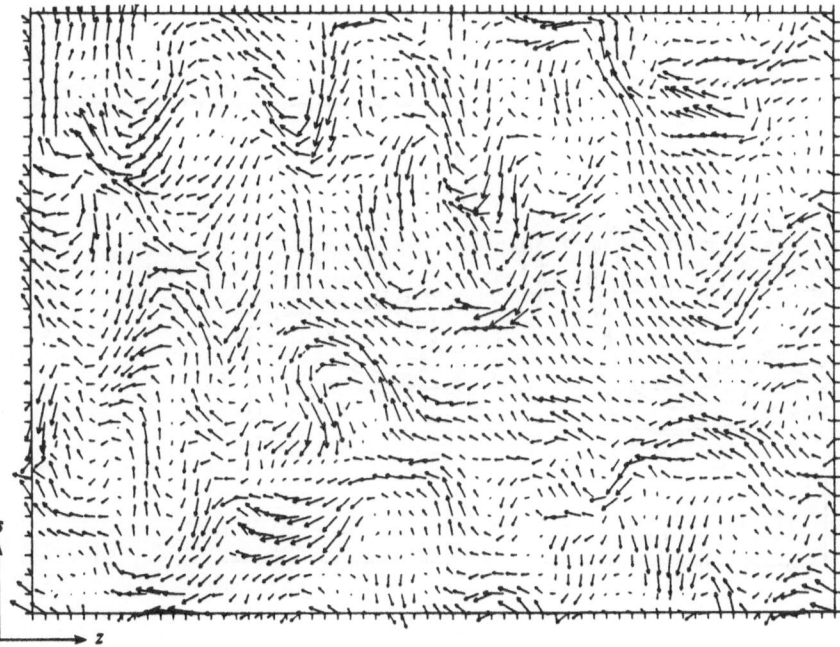

FIG. 5. Homogeneous shear flow. Vorticity vectors in a 45-degree plane[18] (courtesy of Journal of Fluid Mechanics).

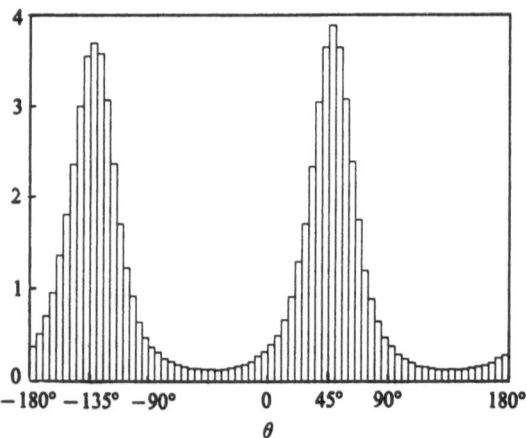

FIG. 6. Homogeneous shear flow. Histogram of the vorticity-vector inclination angles[18] (courtesy of Journal of Fluid Mechanics).

(a)

(c)

(b)

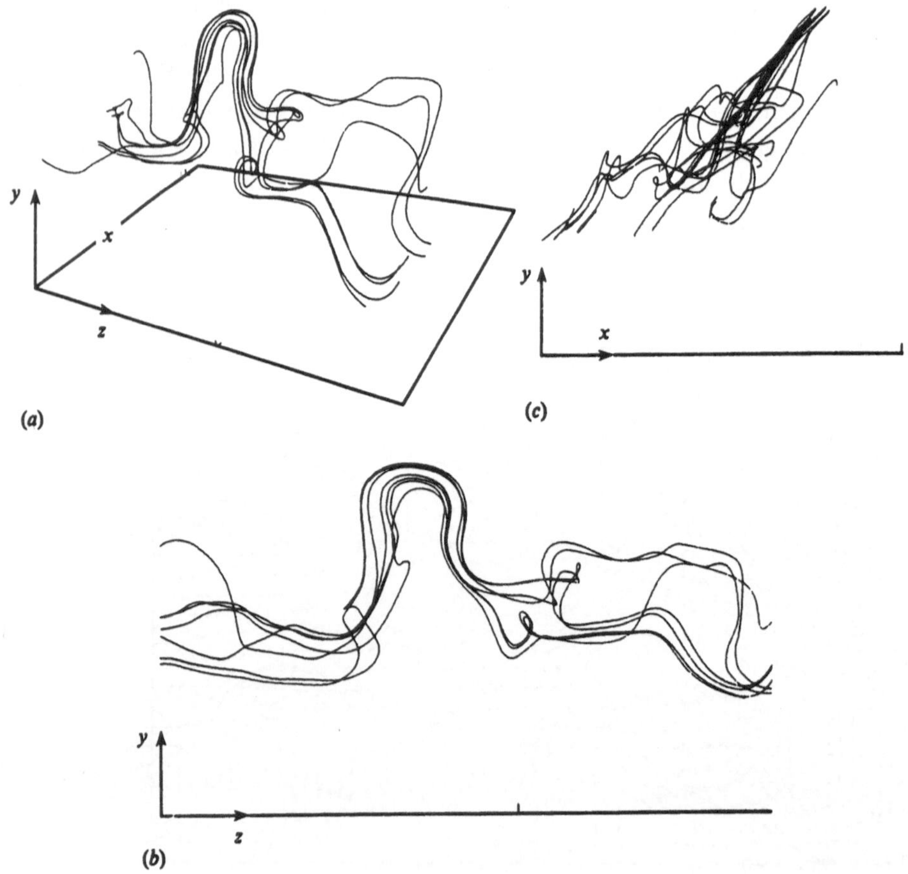

FIG. 7. Homogeneous shear flow. Typical vortex lines displaying a hairpin-like structure[18] (courtesy of Journal of Fluid Mechanics): (a) 3-D view; (b) end view; (c) side view.

We can do other useful things with simulations of this kind. We can locate vortex filaments, even though the computation is grid-based, and not a vortex-tube calculation, by integrating along vortex lines. Then we can visualize the structures. Shown in Fig. 7 are three views of a group of vortex lines displaying a hairpin-like structure in a homogeneous shear flow.

IV. Transitional pipe flow (AL)

A curious phenomenon can exist in pipe flow in a narrow range of Reynolds numbers around $Re = 2200$. Once generated by a sufficiently strong upstream perturbation, long turbulent structures surrounded by laminar flow travel down the pipe without growing or decaying[19]. Lindgren[20] has produced elegant flow visualizations of these so-called turbulent puffs.

In an attempt to unravel some of the mysteries of these objects, we began a program to study transitional pipe flow by numerical simulation. Some of the questions we hoped to answer were: By what mechanisms is the turbulence maintained? Is three-dimensionality important? What vortical structures dominate? As in the case of homogeneous turbulence, the application of spectral methods seemed again to be a good choice for this relatively low-Reynolds-number flow in a relatively simple geometry. We used expansion vectors that were divergence-free and satisfied the no-slip condition at the wall.[21] Thus, the velocity field was assumed to have the expansion

$$ u(r, \theta, x, t) = \sum_{n,k,l} a_{n,k,l}^{(\pm)}(t) \, \chi_n^{(\pm)}(r) \exp(ikx + il\theta) , \qquad (4) $$

where the $(+)$ and $(-)$ represent the two independent vectors per mesh point required for incompressible flow. The determination of suitable basis functions for the radial coordinate was not a trivial matter. We wanted the χ_n's to satisfy the no-slip condition and the divergence condition, to have the correct analytical behavior at $r = 0$, and to allow efficient computations. After a long search, we found that combinations of Jacobi polynomials $P^{(0,\,l)}(2r^2-1)$ and their derivatives worked well and were, in fact, the best choice by far. In particular, the Laplacian operator becomes equivalent to a tridiagonal matrix.

Note that u is periodic in θ, which is an obvious requirement, but it is also assumed to be periodic in x. Don Coles originally suggested the turbulent puff as a candidate for numerical simulation because puffs may be generated experimentally as a periodic train of disturbances, each one separated from its neighbors by laminar flow. Thus periodic boundary conditions along the axis of the pipe are an appropriate choice.

In our first series of computations, the development in time of a number of nonlinear axisymmetric disturbances was investigated. All decayed in relatively short order to a parabolic Poiseuille flow. But a particularly interesting transient structure evolved during one run in which the initial condition consisted of a counter-rotating vortex ring; i.e., a ring containing vorticity opposite in sign to that of the parabolic flow. This structure had many of the features of a turbulent puff; sharp trailing edge, conical leading edge, and a speed of about $0.9U$ (see Fig. 8). These features are consistent with the observation that the disturbance includes vorticity shed from the counter-rotating vortex into its wake as it attempts to swim upstream. The shed vorticity moves downstream with the local axial velocity and decays; hence the conical leading edge.

In an attempt to obtain a self-sustaining, puff-like structure, a three-dimensional simulation was initiated with the axisymmetric, counter-rotating structure of Fig. 8 but

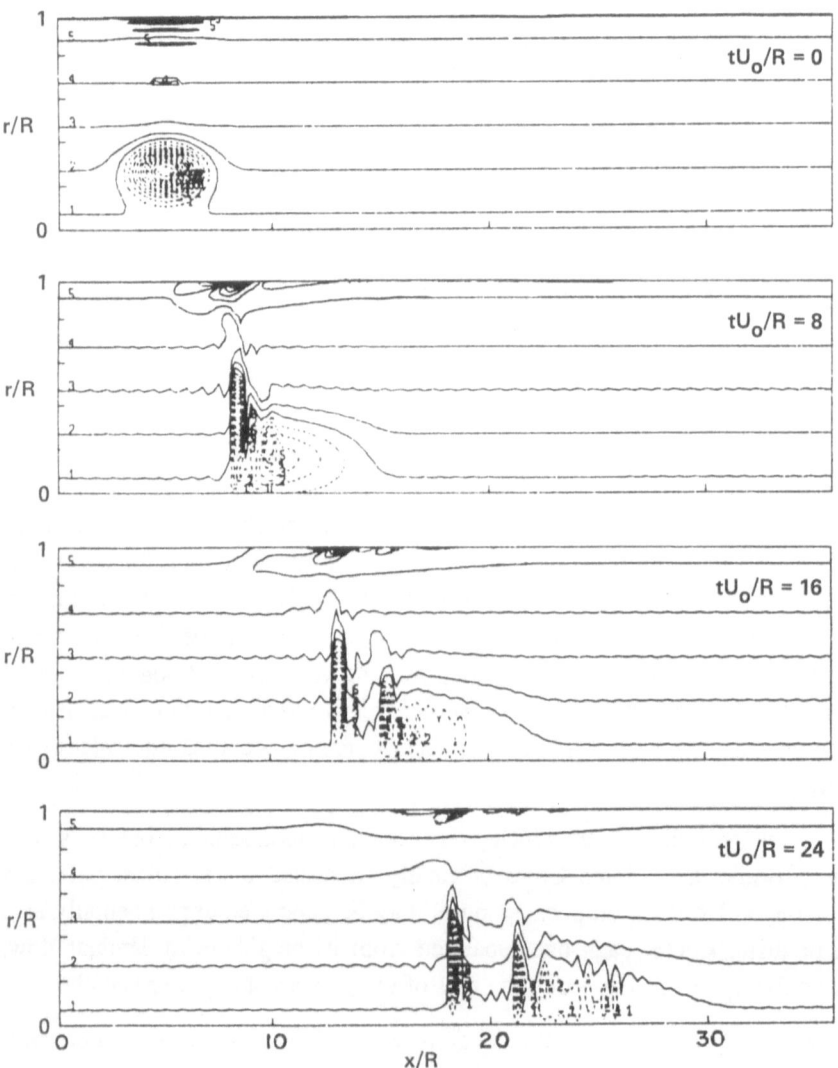

FIG. 8. Evolution of a counter-rotating, axisymmetric vortex ring in a pipe. $Re = 2200$. Contours of $\omega_\theta(r,x)$.

with a low-level background of random, three-dimensional perturbations. The results, shown in Fig. 9, are very suggestive of a turbulent puff. The number of Fourier modes was $N_k = 240$ and $N_l = 32$, and the number of radial polynomials was $N_n = 49$, for a total of approximately 0.4×10^6 points in spectral space. The number of points in real space was 1.5×10^6, the number required to eliminate aliasing. Even so, the axial dimension of the periodic domain had to be restricted to only 18 diameters, somewhat short of the 40 diameters or so that is characteristic of laboratory puffs. The simulation was run on the Cray X-MP/12 at NASA Ames. The program was written in the VECTORAL language of Alan Wray.

As can be seen in Fig. 10, during the early stages of development, $t < 20$, the total energy of the disturbance decays fairly rapidly, but the axisymmetric structure is unstable to three-dimensional perturbations, so that the non-axisymmetric energy grows rapidly during this time. The disturbance seems nearly to settle down as a quasi-steady three-dimensional object for $60 < t < 80$, but then roughly three more periods of growth are experienced, each one less vigorous than the previous one. The non-axisymmetric energy peaks during the first of these additional growth periods but then levels off.

Although the analysis of this simulation, which took about four hundred hours of cpu time over a period of ten months, is far from complete, we can make the following preliminary observations:

1. *Three-dimensionality is vital to the maintenance of the puff.* Recall that all axisymmetric disturbances that were tried decayed initially. To further test this hypothesis, we can take advantage of the fact that we are performing numerical experiments and restart the three-dimensional velocity field at $t = 80$ with all non-axisymmetric modes $(l \neq 0)$ set to zero. The result, shown in Fig. 11, should be compared with the three-dimensional result shown in Fig. 9. It is seen that the axisymmetric continuation in time decays rapidly to Poiseuille flow.

2. *A mechanism necessary to the maintenance of the turbulent puff is the passage of co-rotating vortex tubes through the axis to become counter-rotating vortex structures.* The argument is as follows. A feature of all turbulent puffs is a deficit in streamwise velocity along the axis; see Fig. 12, where we show a typical trace from the simulation and one from experiment. The total integrated deficit is equal to the circulation of the disturbance,

$$\int u_x \, dx = \iint \omega_\theta \, dr \, dx = \Omega \ . \tag{5}$$

By integrating the transport equation for the disturbance vorticity, we find that

$$\frac{d\Omega}{dt} = \int (\omega_\theta' u_r' - u_\theta' \omega_r')_{r=0} \, dx + \text{viscous terms} \ , \tag{6}$$

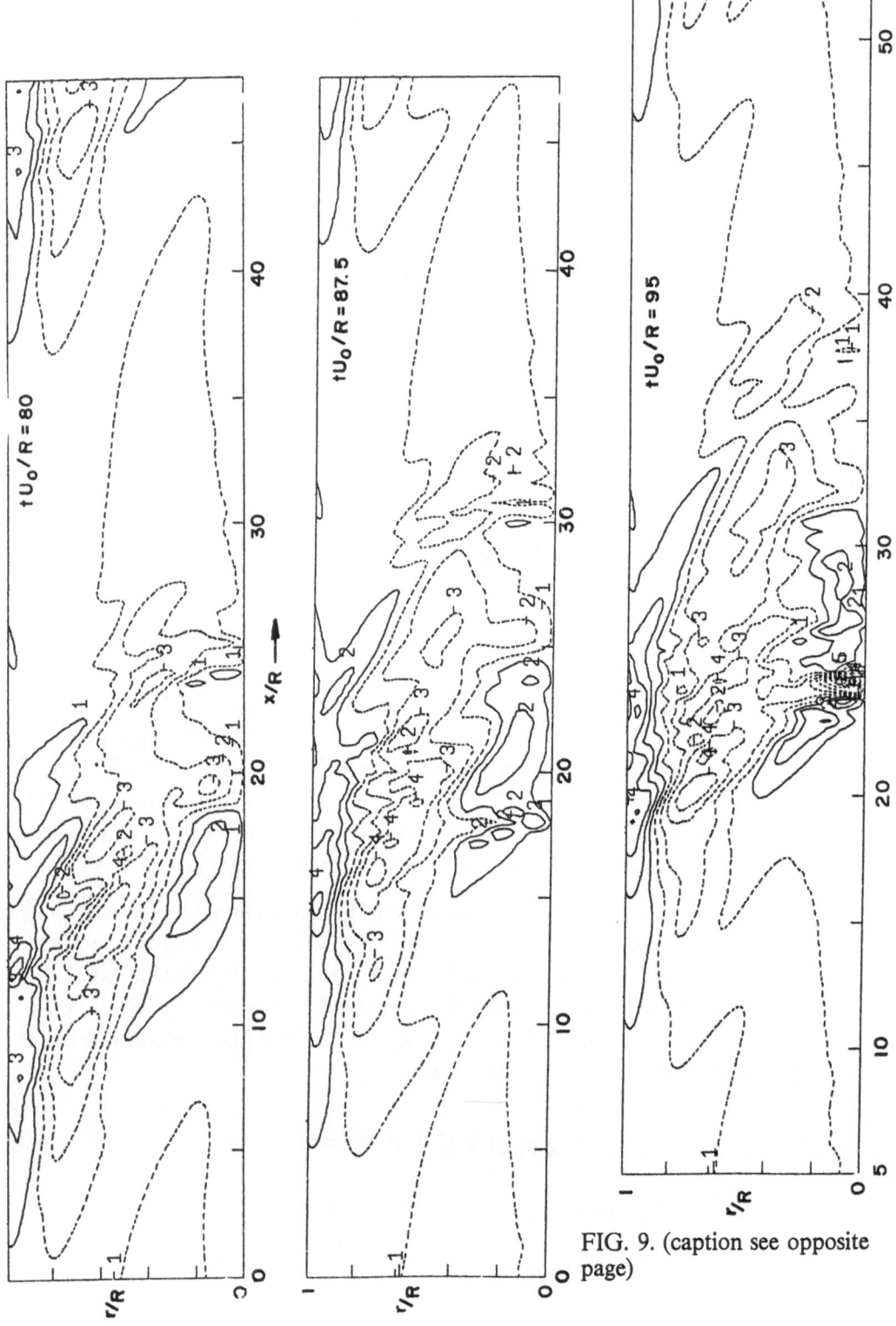

FIG. 9. (caption see opposite page)

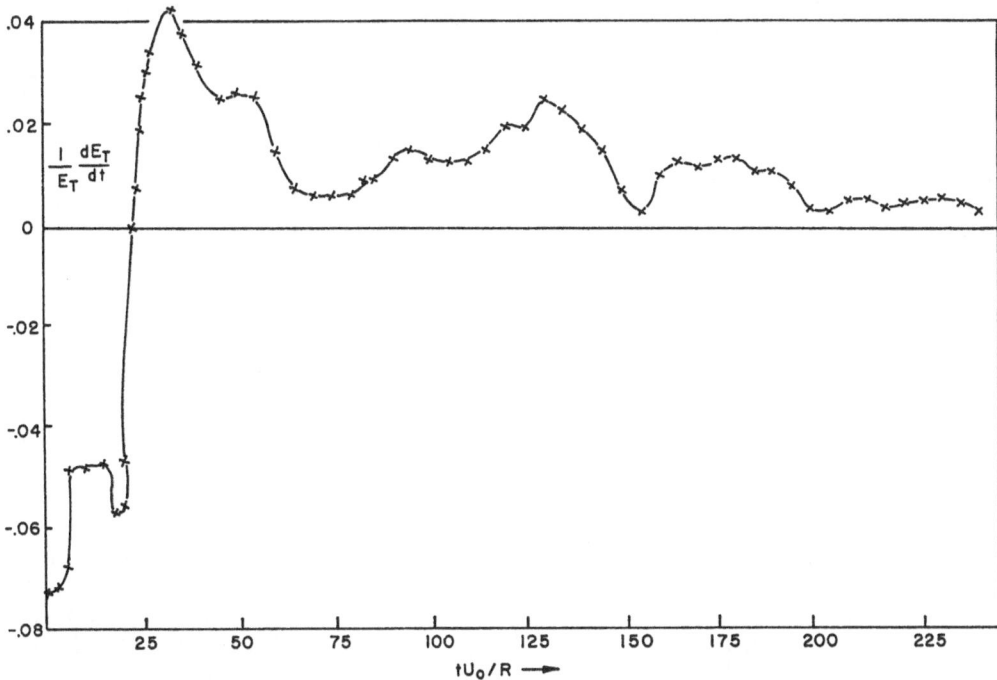

FIG. 10. Growth rate of the three-dimensional disturbance in a pipe.

where the primed quantities denote non-axisymmetric components. By kinematics, the primed terms above must come from components $l = \pm 1$ and represent the flux of ω_θ' (or, equivalently, ω_r') vorticity across the $r = 0$ axis. We see evidence for activity of this sort in Fig. 9 at $t = 87.5$ and $t = 95$ at the rear of the puff on the axis. We suspect that there is a significant inviscid contribution to $d\Omega/dt$; i.e., a co-rotating vortex tube is starting to pass through the origin. This suspicion has been confirmed by studying contour plots of ω_θ ($l = 1$) and u_r ($l = 1$) and computing the change in Ω during this time, $87.5 < t < 95$. The inviscid contribution to Ω was negative, while the viscous contribution to Ω was positive and about one-third the magnitude of the inviscid contribution during the same period of time.

We have seen that a train of vortex rings in a pipe, when subjected to three-dimensional disturbances, can lead to a train of turbulent puffs. Next we shall see what happens for the case of a train of vortex rings in free space; i.e., an excited round jet. The results are quite amazing.

FIG. 9. Snapshots of the three-dimensional disturbance in a pipe. $Re = 2200$. Contours of

$$\overline{\omega}_\theta(r,x) = \frac{1}{2\pi} \int_0^{2\pi} \omega_\theta(r,x,\theta) \, d\theta \ .$$

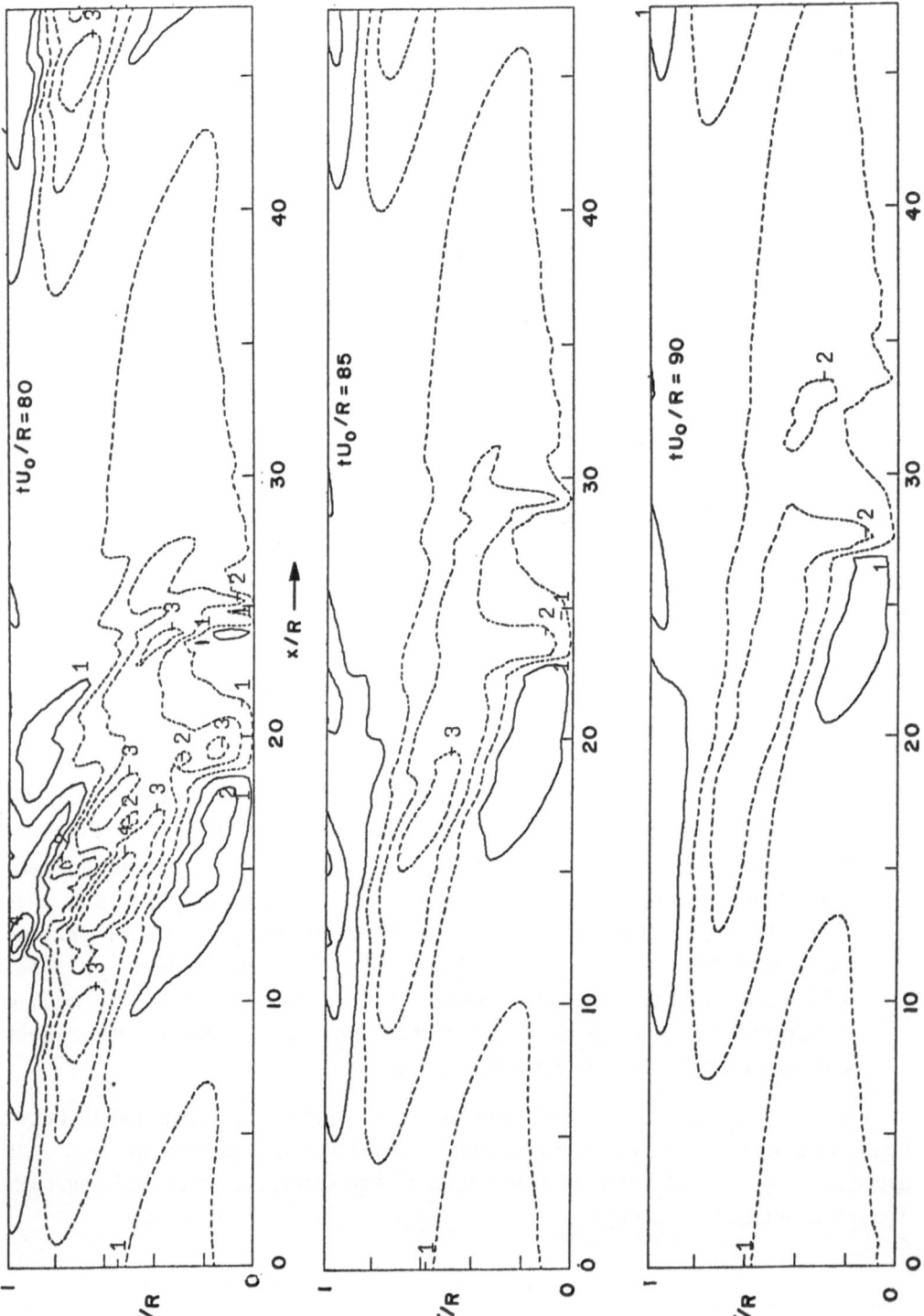

FIG. 11. Evolution of a purely axisymmetric perturbation. Initial condition is derived from the three-dimensional disturbance at top of Fig. 9.

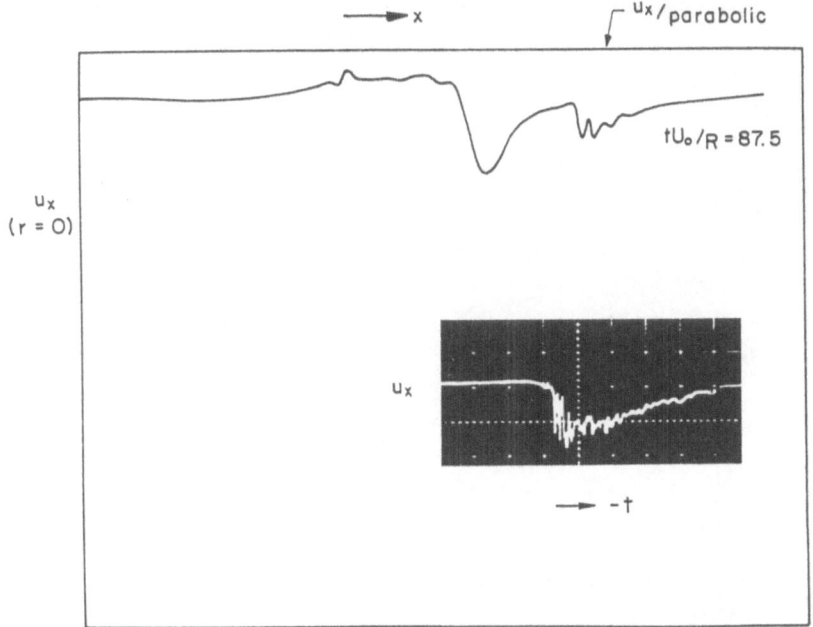

FIG. 12. Axial velocity along the pipe center for the three-dimensional disturbance. Inset: experiment of Ref. 19 (courtesy of I. Wygnanski).

V. Excited round jets (WCR)

My latest favorite subject is the bifurcating and blooming jet[22]. Figure 13 is a picture taken when I was here at Caltech in 1983-84. This jet has a Reynolds number of the order of 8000 and is being excited by axial disturbances and also by orbital disturbances caused by wiggling the nozzle very, very slightly. There is a about a 15-percent amplitude in the orbital excitation and a few percent amplitude in the axial excitation. Under certain conditions, explained below, the jet bifurcates and can be made to spread with a large spreading angle. The jet actually becomes two jets, and if we look at the flow far from the origin we can convince ourselves that the two legs of the jet are diverging faster than either leg is spreading. It follows that the far field of this jet can never become the far field of a normal jet, in spite of "principles" that claim otherwise. Figure 14 is an axial view of the same flow. Evidently the sidewise spreading is normal; it is the transverse spreading that is abnormal.

Another possible outcome with this sort of exitation is the blooming jet shown in Figs. 15 and 16. Basically, the jet is composed of a shower of vortex rings. With the right kind of excitation, this jet can have a spreading angle as large as 135 degrees. The one shown has almost a 90-degree spreading angle. There is tremendous mixing in the jet plume. We have measured the velocity decay in this jet, and found it to be exponential. So it is a very rapidly decaying and mixing jet.

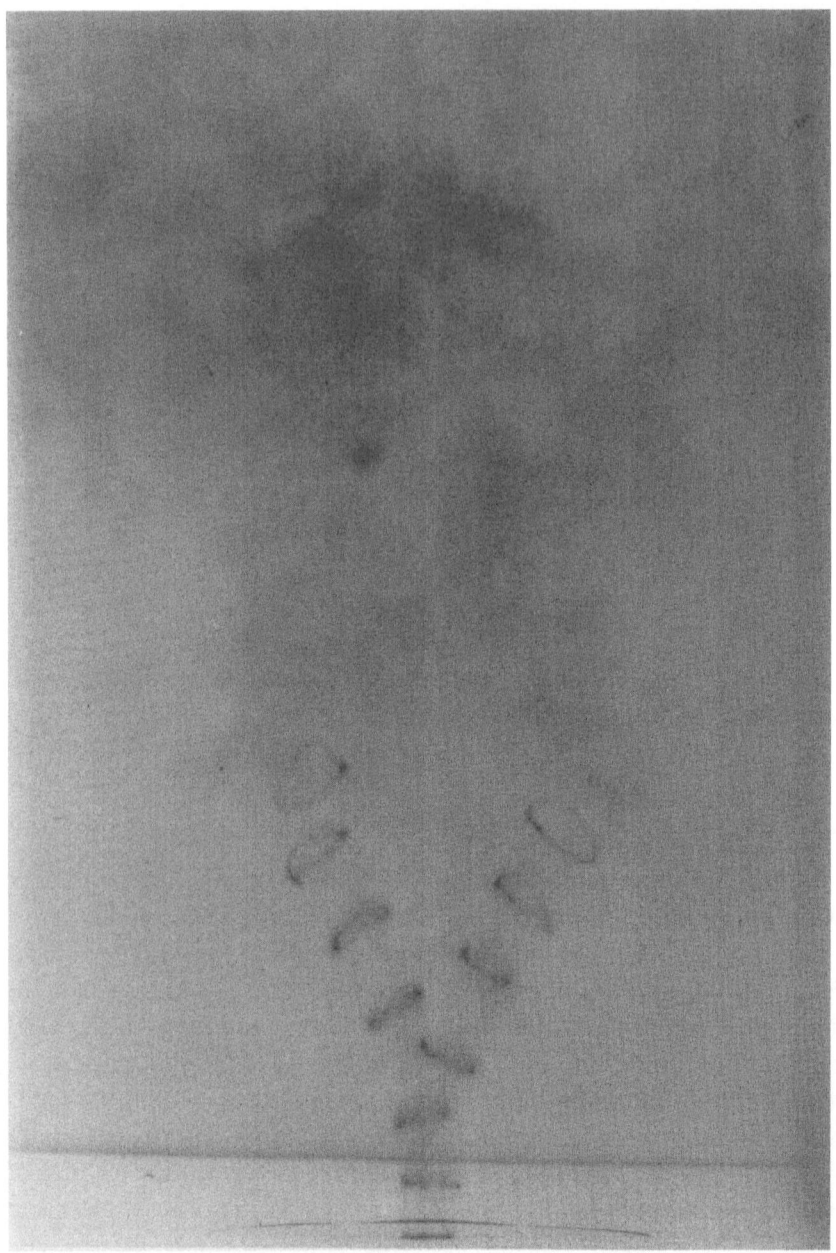

FIG. 13. Side view of bifurcating jet[22] showing two diverging legs.

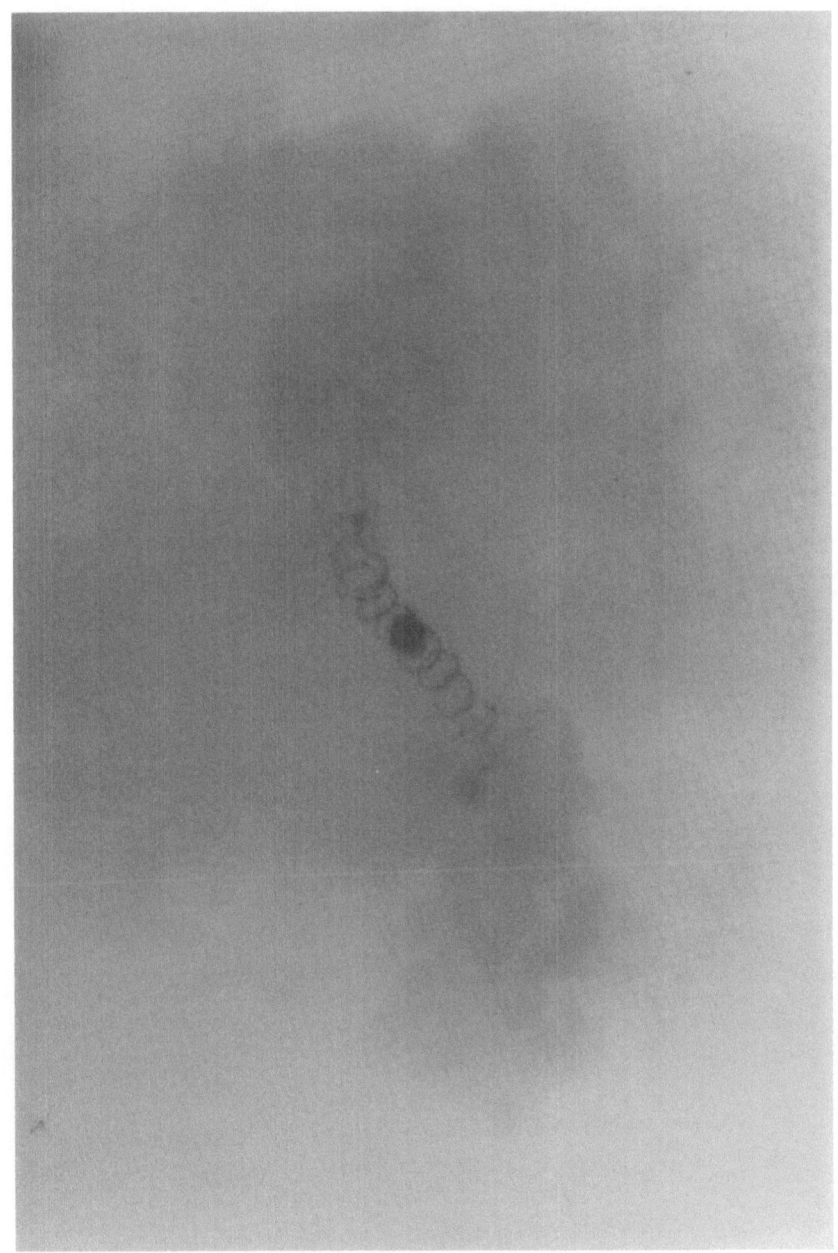

FIG. 14. Axial view of bifurcating jet[22].

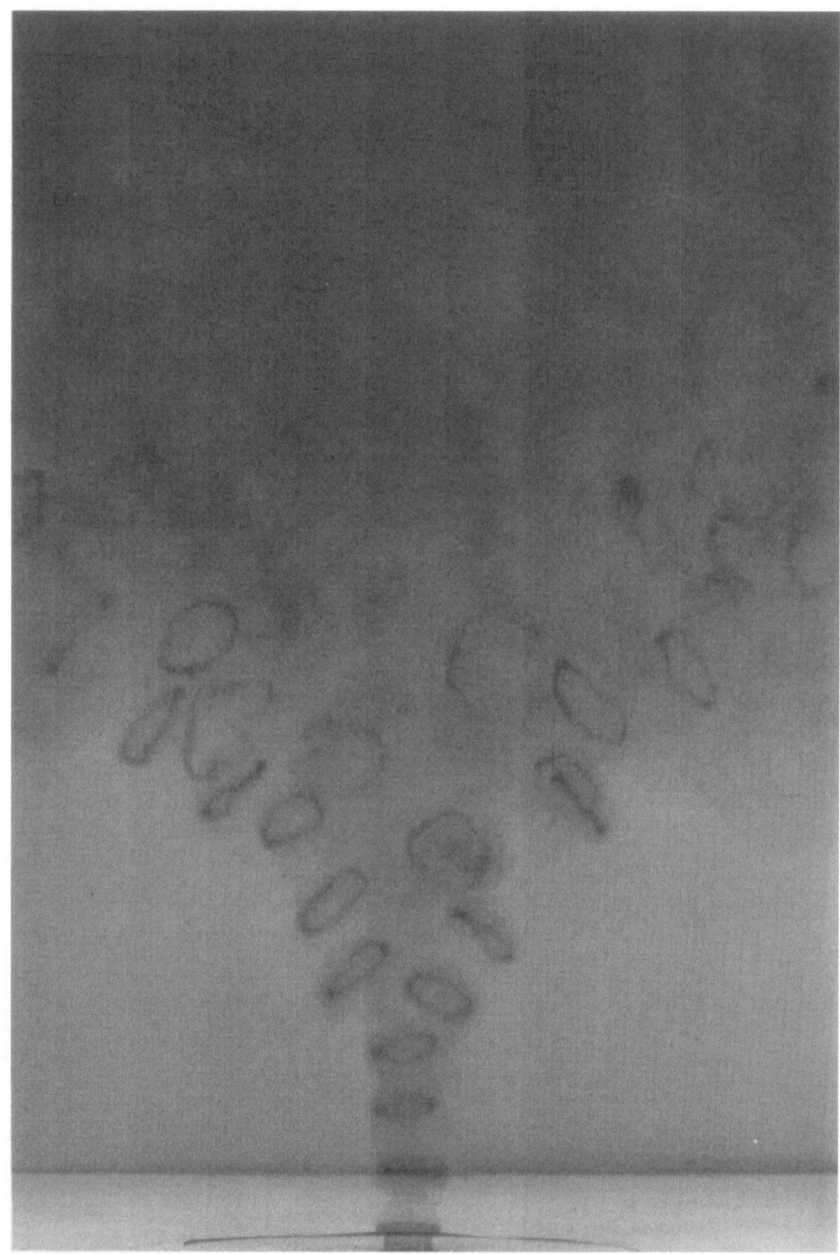

FIG. 15. Side view of blooming jet[22].

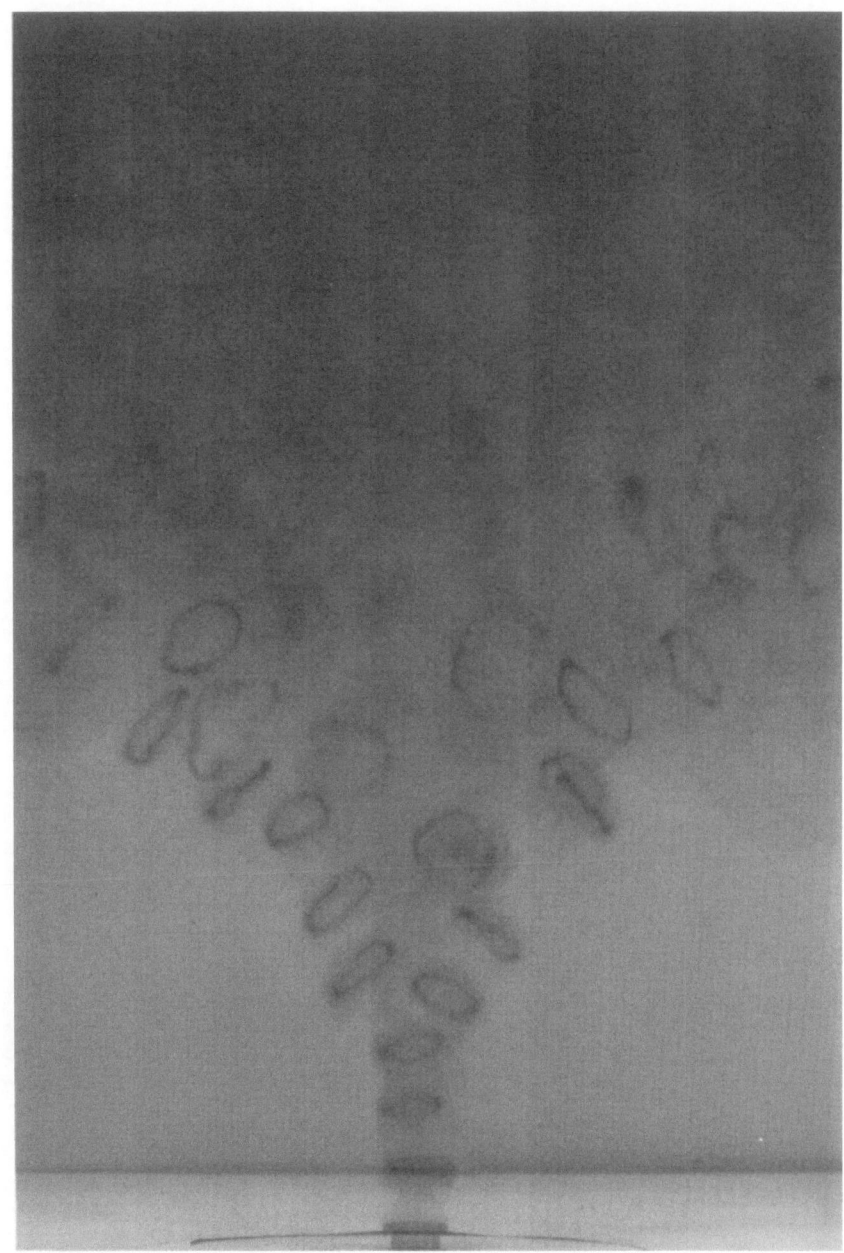

FIG. 15. Side view of blooming jet[22].

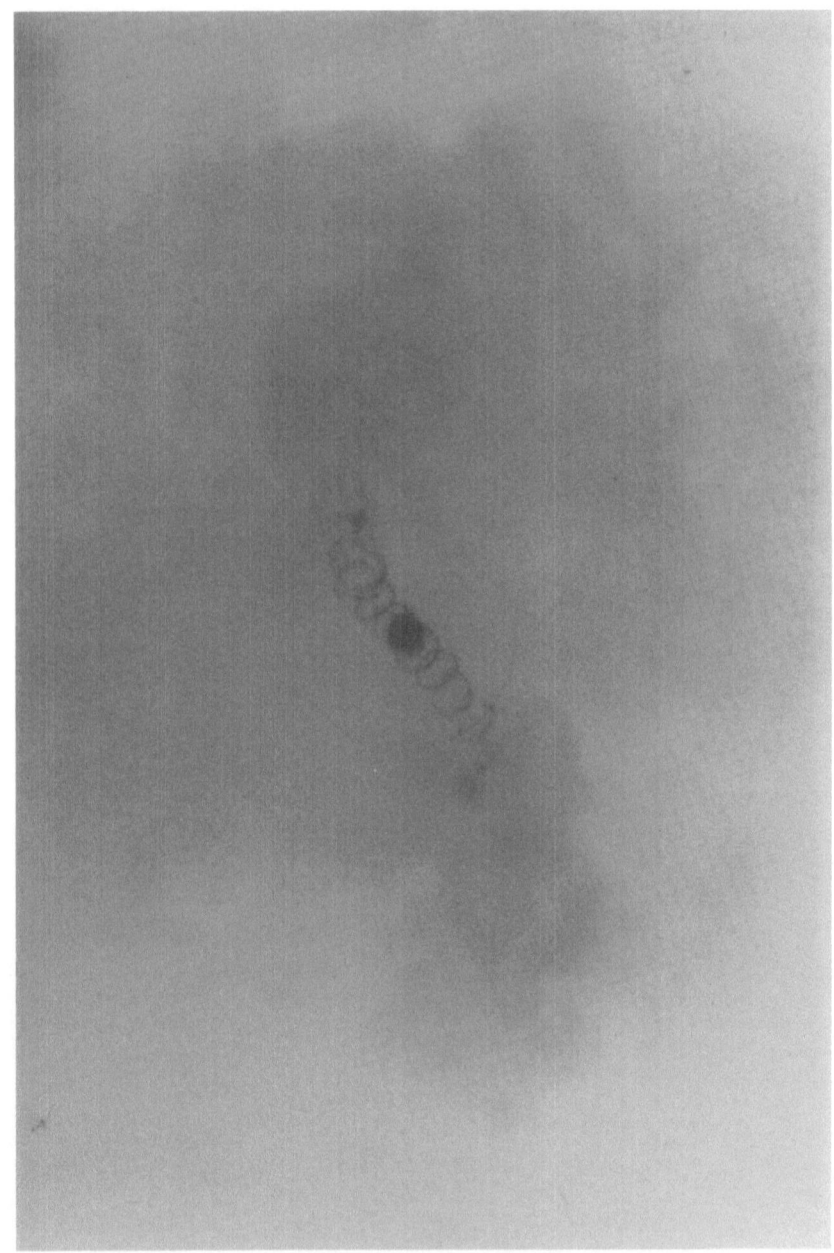

FIG. 14. Axial view of bifurcating jet[22].

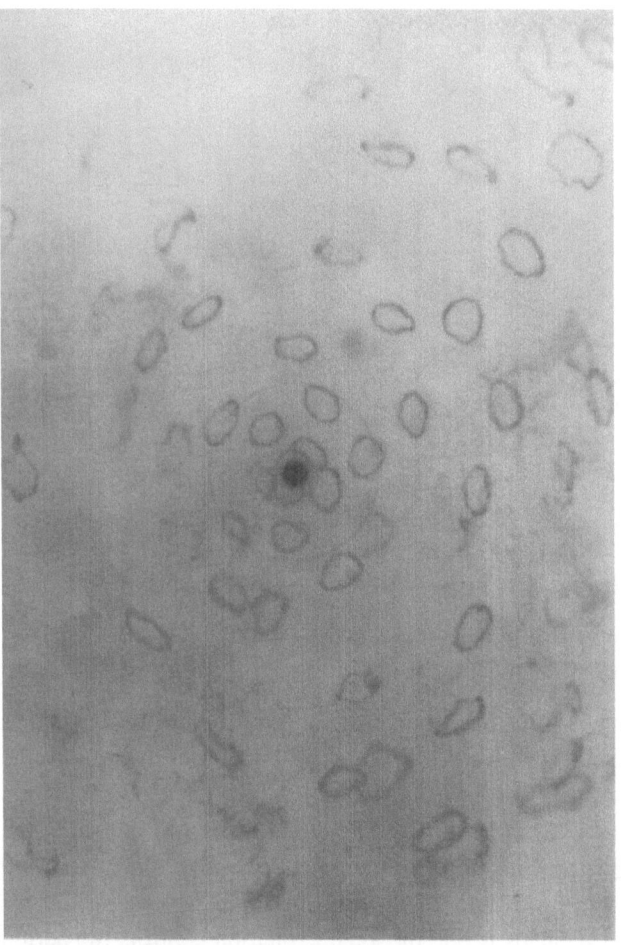

FIG. 16. Axial view of blooming jet[22].

We can also study and try to understand such jets by using numerical simulations based on vortex filaments. A student, David Parekh, and I enlisted Tony to assist us with the simulations. We began by mapping out the regimes in which we could observe these different effects. We generate a bifurcating jet when the axial excitation frequency is twice the orbital excitation frequency. Two vortex rings are shed per orbit, and these separate, each going into one of the two downstream legs of the jet. The blooming jet is obtained for the range of parameters shown in Fig. 17. The purpose of the simulations was to understand better the underlying processes. We used three-dimensional vortex filaments to represent the flow, adding a vortex ring whenever one was shed as a result of the axial excitation. The rings are put in slightly eccentrically, so that they are slightly off-axis from one another. What happens? From Fig. 18, which displays the results of the simulation, it is apparent that the vortex rings are indeed tilting, as in the laboratory experiment.

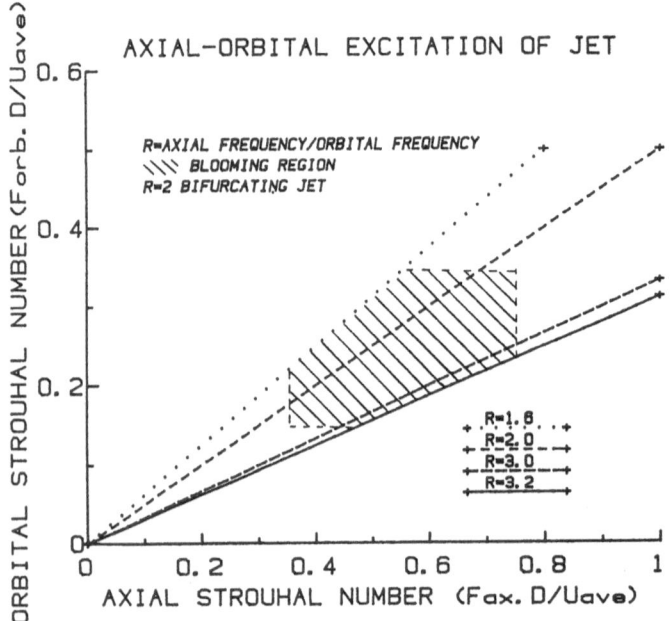

FIG. 17. Regimes of the perturbed round jet as determined by experiment[22].

What have we learned? First, we have learned that the controlling mechanism is vortex induction. The phenomena of tilting and splitting were discovered in the laboratory, but the explanation was provided by the simulation. Second, the simulation predicted that if we keep the ratio of frequencies fixed at two, but increase the frequency, then the spreading angle will increase. We predict spreading angles up to 50 degrees, as shown in Fig. 19. Then we reach a critical condition beyond which we cannot get a bifurcating jet. What happens is that the axial excitation controls the period of the rings, and the orbital excitation controls their eccentricity. So if we increase the frequencies but keep the ratio equal to two, the rings move closer together. The closer together they are, the more they tilt each other; but if they are too close to each other, they become tangled before they can separate. This result was discovered by the numerical simulation and subsequently confirmed in experiments both at Caltech and at Stanford.

JET EXIT

FIG. 18. Three-dimensional vortex simulation of the bifurcating jet by Parekh.

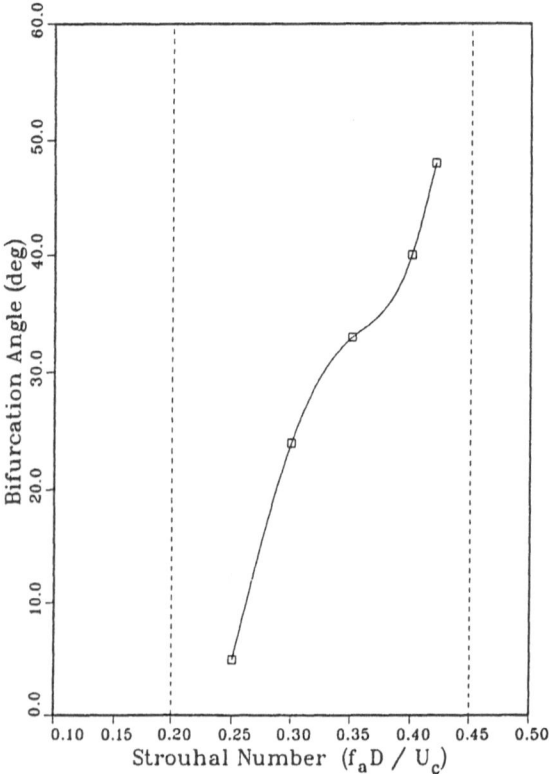

FIG. 19. Spreading angle of the bifurcating jet as a function of Strouhal number, from Parekh's numerical simulation.

VI. More wall-bounded flows (WCR)

We return now to the area of full simulations of wall-bounded flows. Tony will describe the recent work of Spalart on turbulent boundary layers[23], and I will describe Moser's simulations of curved channel flow[24]. But we are running out of time, so we will have to go to parallel sessions.

WCR: Moser studied a curved channel flow, using a full simulation of all of the important scales of turbulence. The channel flow is attractive because periodic boundary conditions can be imposed in the flow direction. Even so, the development of an efficient spectral method for this curved geometry was a non-trivial task.

AL: Spalart had an idea for dealing with non-periodic boundary conditions in the flow direction, and he wanted to try this idea out on a spatially-developing boundary layer. He also had to develop a spectral method to deal with the semi-infinite geometry.

WCR: In the large-eddy simulations (LES) of Moin and Kim[25], the wall-layer structures were too large, and Moser wanted to see if the right scales would emerge from a full simulation.

AL: Moser did establish that a full simulation would give the right scales in the wall region, and Spalart wanted to see if it would in the boundary layer, too.

WCR: Here are the mean profiles (Fig. 20)...

AL: ... and here are the turbulence profiles (Fig. 21).

WCR: Note that the mean profiles are in excellent agreement with experiments ...

AL: ... as are the turbulence profiles.

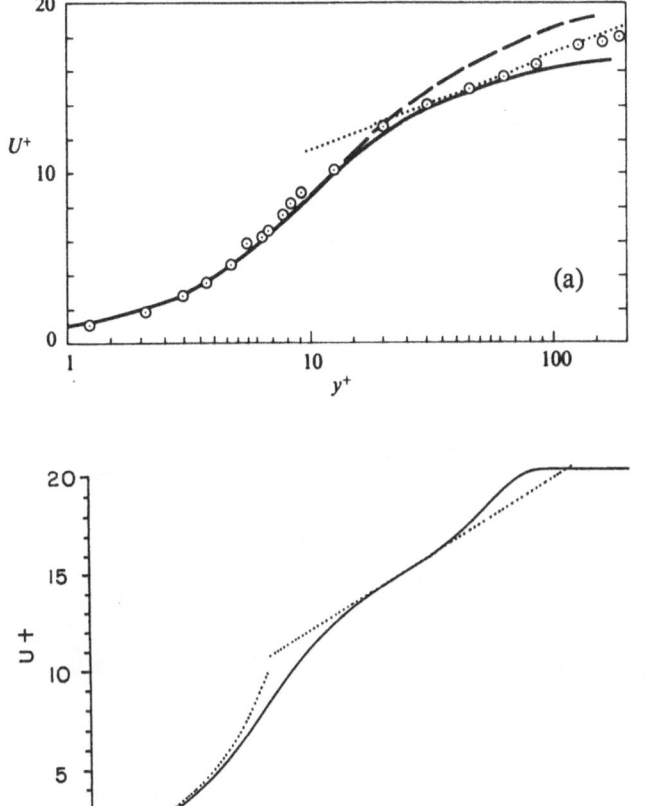

FIG. 20. Full simulation of wall-bounded flows. Mean profiles for: (a) curved channel flow[24] (courtesy of Journal of Fluid Mechanics); (b) turbulent boundary layer[23]. Experimental data in (a) from Ref. 30.

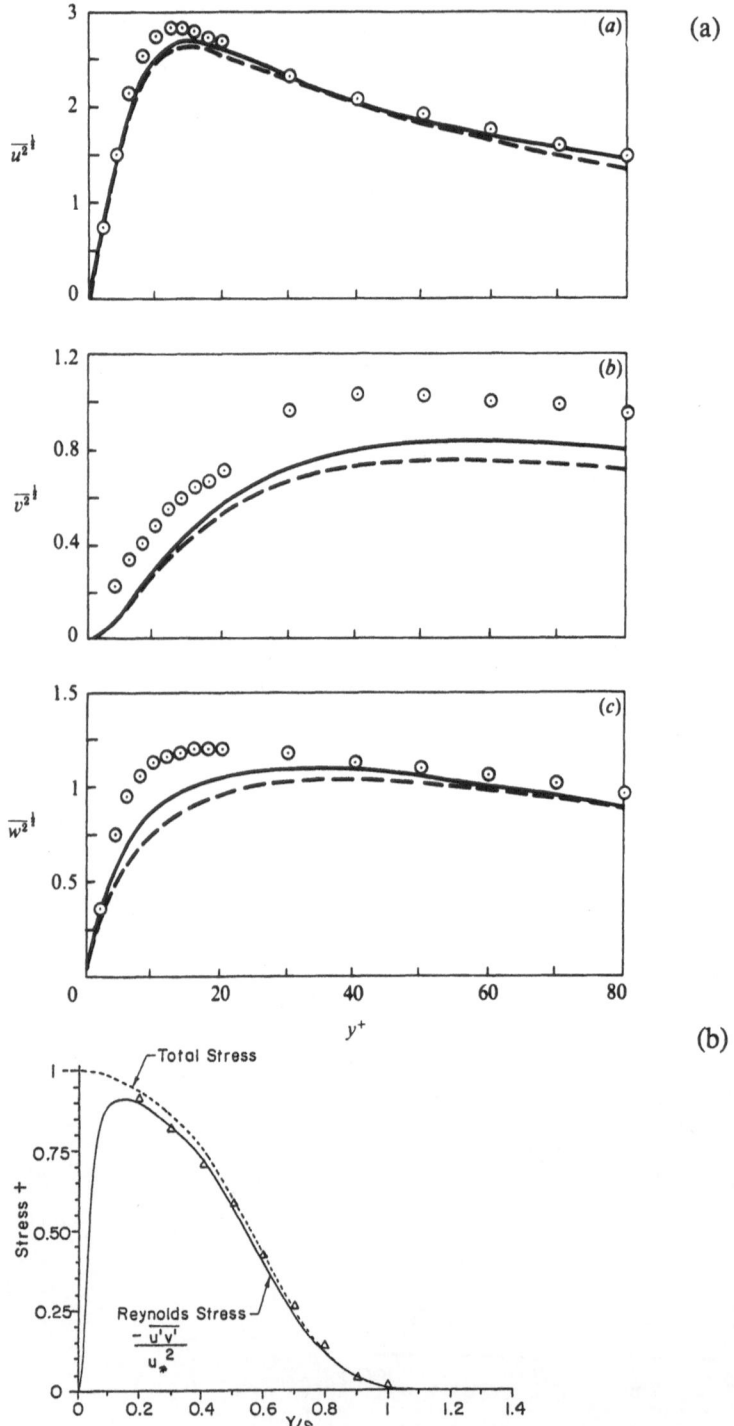

(a)

(b)

FIG. 21. Full simulation of wall-bounded flows. Turbulence profiles for: (a) curved channel flow[24] (courtesy of Journal of Fluid Mechanics); (b) turbulent boundary layer[23]. Experimental data in (a) from Ref. 31; in (b) from Ref. 32.

WCR: Moser found the Görtler vortices that are seen in experiments, and was able to isolate them in his data. He found that they contribute as much as forty percent of the Reynolds shearing stress, which is why standard models of turbulence fail so badly in curved boundary-layer flows.

AL: Spalart is not as far along in his work, but he has been examining an old debate as to whether the Reynolds stress $\overline{u'v'}$ varies as y^3 or y^4 at the wall. After he answers this question, he wants to explore the near-wall structure in boundary layers as a function of Reynolds number and pressure gradient. Spalart has completed a series of these calculations[23] and has made a number of curious discoveries. Each one, however, leads to new questions that should be answered. Two examples are: (1) a log layer in the sink boundary layer that has an unexpectedly large extent, and (2) a two-dimensional mixing-layer-like character in the outer part of Stratford's separating boundary layer. Moreover, Spalart has found that turbulence profiles near the wall and in the wake region have a complex dependence on Reynolds number, indicating an urgent need for improved scaling laws.

WCR: Moser wants to identify structures in his flow, as Moin and Kim have done recently with their LES fields. He also wants to see if he can identify strange attractors in his flow, and perhaps use his fields as a way to evaluate some new ideas about chaos and turbulence.

AL: It is quite clear to Bill and me that this sort of full turbulence simulation will be an important new way to study the physics of turbulence. Rapid advances in three-dimensional graphics make this approach especially attractive. A turbulence researcher can now sit at a three-dimensional animated display console, like the one Hesselink is pioneering at Stanford, or at a powerful graphics work-station with special software, like the software written by Buning[26] at NASA Ames, and view vortex filaments, pressure peaks, and so on, from all sorts of different angles, in an effort to gain understanding of the flow.

WCR: Finally, we will conclude our discussion with some remarks about computations of flows of practical interest. For the first of these we have chosen a simulation of the time-dependent flow through a set of acoustic baffles proposed for the inlet of the 80-foot by 120-foot test section at NASA's Ames Research Center. These calculations were designed to find out if the pressure losses would be acceptable.

VII. Separated flows (AL)

We come now to our final topic -- flows with large-scale, unsteady separation at high Reynolds numbers. We will confine our discussion to flows in two dimensions, where a number of reasonably successful simulations have been achieved. (A number of failures have been achieved also.) Although these flows are two-dimensional, some are of considerable engineering interest. Needless to say, when reliable computational methods become available for three-dimensional geometries, interest in these methods should increase dramatically.

For constant-density flows past solid bodies, we observe that vorticity is generated only at the no-slip boundaries of solid surfaces. At high Reynolds numbers, once the vorticity separates from the boundary layer, it moves with the fluid with very little diffusion. This observation suggests that we again use a vortex method, in which parcels of vorticity move with the local fluid velocity, to simulate the dynamics of the wake. But two important elements must be added to complete the picture:

1. Satisfy the inviscid boundary condition at the solid surface; i.e., no flow through the walls, and

2. Determine the locations of the separation points in time and the flux of vorticity into the outer flow.

The first element is rather easy to take care of. We can simply use a boundary integral method, one form of which is the panel method, where vortex tiles are laid along the boundary and the circulation of each tile is determined from the solution of a linear system representing the mutual influences of the tiles. As an alternative procedure for simple geometries (i.e., any shape that can be transformed to a circle by conformal mapping), the method of images may be used. The second element is, in general, much more difficult. Here we have to compute the unsteady mechanics of the separating boundary layer. If we want to simulate a flow having a turbulent boundary layer, we are limited to available models for these flows. However, if the boundary layer is laminar, we should in principle be able to compute what is required. This computation has been done at several levels of sophistication with considerable success[27,28,29], but more work needs to be done.

Figure 22 illustrates an application of the method by Spalart to predict the total-pressure loss for flow past a set of acoustic baffles proposed to be located upstream of the inlet to the 80-foot by 120-foot test section. It had been estimated previously that the design shown would lead to an acceptable pressure-loss coefficient of approximately 5 to 10. Spalart's simulation predicts a much higher loss coefficient of 86. This pessimistic result was confirmed later in an experimental study by John Foss, who found a value of 82.

Unfortunately, such simulations do not always agree with the corresponding experiments on two-dimensional geometries. For example, force coefficients may be off by as much as fifty percent. We find, however, that if there exists some phenomenon that promotes two-dimensionality or spanwise coherence in the experiment, such as an oscillation of the body, then the agreement tends to be much better. Thus, three-dimensionality of the flow past a two-dimensional object may well be to blame when substantial disagreement is observed.

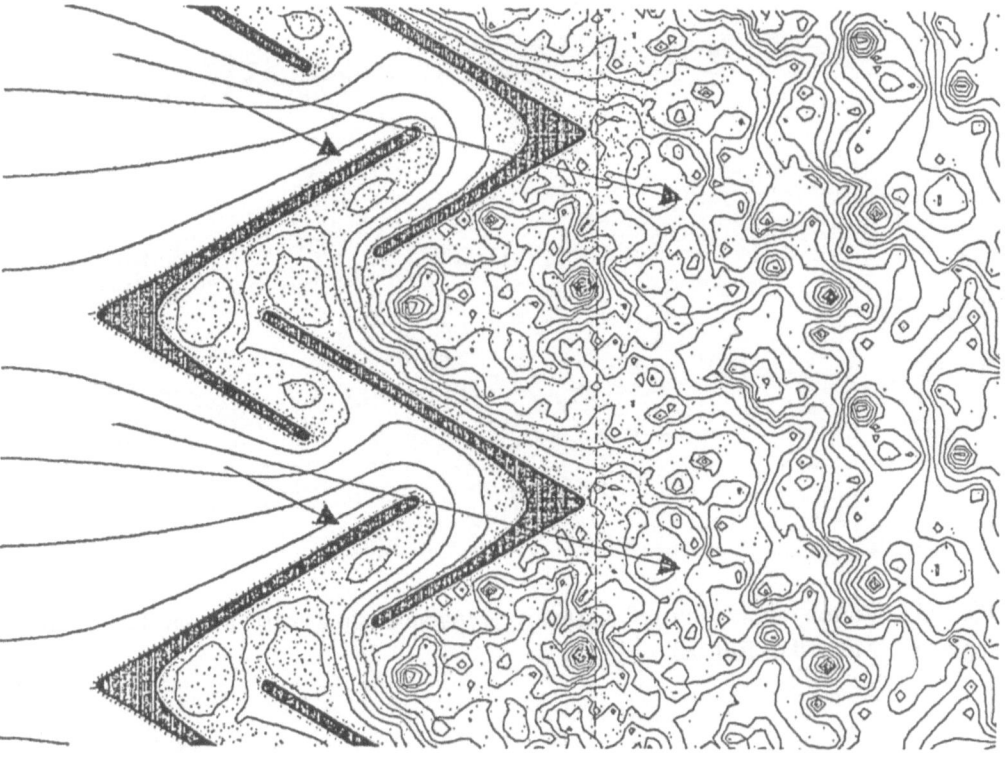

FIG. 22. Vortex simulation of two-dimensional flow through a set of acoustic baffles. Method of Ref. 28.

Our final example is designed to be of immediate use to H. W. Liepmann, perhaps during a sailing excursion with Paul Dimotakis, and is illustrated in Fig. 23. Perhaps the real three-dimensional Liepmann will discount our primitive attempt to simulate such a complex flow, but we hope that he will applaud our efforts and encourage us to do better next time.

FIG. 23. Vortex simulation of unsteady, separated flow past a two-dimensional
Liepmann. Method of Ref. 28.

References

1. K. Shariff, J.H. Ferziger, and A. Leonard, SIAM Fall Meeting, Phoenix, Arizona, 1985.

2. C. Pozrikidis, J. Fluid Mech. **168**, 337 (1986).

3. M.J. Zabusky, M.H. Hughes, and K.V. Roberts, J. Comput. Phys. **30**, 96 (1979).

4. D.G. Dritchel, J. Fluid Mech. **172**, 157 (1986).

5. P.G. Saffman and R. Szeto, Phys. Fluids **23**, 2339 (1980).

6. R.T. Pierrehumbert, J. Fluid Mech. **99**, 129 (1980).

7. J. Norbury, J. Fluid Mech. **57**, 417 (1973).

8. Y. Oshima, J. Phys. Soc. Japan **44**, 328 (1978).

9. K. Shariff, A. Leonard, N. Zabusky, and J.H. Ferziger, Fluid Dyn. Res. **3** (to appear, 1988).

10. W. Mohring, J. Fluid Mech. **85**, 685 (1978).

11. T. Kambe and T. Minota, Proc. R. Soc. London **A386**, 277 (1983).

12. R.A. Wigeland and H.M. Nagib, IIT Fluid and Heat Transfer Report R78-1, Illinois Institute of Technology, Chicago, Illinois, 1978.

13. J. Bardina, J.H. Ferziger, and R.S. Rogallo, J. Fluid Mech. **154**, 321 (1985).

14. E. Shirani, J.H. Ferziger, and W.C. Reynolds, Department of Mechanical Engineering, Report TF-15, Stanford University, Stanford, California, 1981.

15. C.-T. Wu, J.H. Ferziger, and D.R. Chapman, Department of Mechanical Engineering, Report TF-21, Stanford University, Stanford, California, 1985.

16. M.J. Lee and W.C. Reynolds, Department of Mechanical Engineering, Report TF-24, Stanford University, Stanford, California, 1985.

17. R. Narasimha, J. Indian Inst. Sci. **64A**, 1 (1983).

18. M.M. Rogers and P. Moin, J. Fluid Mech. **176**, 33 (1987).

19. I.J. Wygnanski and F.H. Champagne, J. Fluid Mech. **59**, 281 (1973). See also I. Wygnanski, M. Sokolov, and D. Friedman, J. Fluid Mech. **69**, 283 (1975).

20. E.R. Lindgren, Ark. Fys. **12**, 1 (1957).

21. A. Leonard and A. Wray, in *Proceedings of the 8th International Conference on Numerical Methods in Fluid Dynamics*, Aachen (Springer-Verlag, New York, 1982).

22. M. Lee and W.C. Reynolds, Department of Mechanical Engineering, Report TF-22, Stanford University, Stanford, California, 1985.

23. P.R. Spalart and A. Leonard, in *Proceedings of the Fifth Symposium on Turbulent Shear Flows*, Ithaca (Springer-Verlag, 1987). See also P.R. Spalart, J. Fluid Mech. **172**, 307 (1986).

24. R.D. Moser and P. Moin, J. Fluid Mech. **175**, 479 (1987). See also R.D. Moser, P. Moin, and A. Leonard, J. Comput. Phys. **52**, 524 (1983).

25. P. Moin and J. Kim, J. Fluid Mech. **118**, 341 (1982).

26. P. Buning (private communication).

27. P.R. Spalart, A. Leonard, and D. Baganoff, NASA Report TM-84328, 1983.

28. P.R. Spalart, AIAA Paper 84-0343, 1984.

29. P.R. Spalart, J. Prop. Power **1**, 235 (1985).

30. J.M. Wallace, H. Eckelmann, and R.S. Brodkey, J. Fluid Mech. **54**, 39 (1972).

31. H.-P. Kreplin and H. Eckelmann, Phys. Fluids **22**, 1233 (1979).

32. P.S. Klebanoff, NACA TN 3178, 1954.

Geologic Nozzles[1]

Susan Werner Kieffer

U.S. Geological Survey, Flagstaff, Arizona 86001

Abstract

Sonic velocities of geologic fluids, such as volcanic magmas and geothermal fluids, can be as low as 1 m/s. Critical velocities in large rivers can be of the order of 1-10 m/s. Because velocities of fluids moving in these settings can exceed these characteristic velocities, sonic and supersonic gas flow and critical and supercritical shallow-water flow can occur. The importance of the low characteristic velocities of geologic fluids has not been widely recognized and, as a result, the importance of supercritical and supersonic flow in geological processes has generally been underestimated. The lateral blast at Mount St. Helens, Washington, propelled a gas heavily laden with dust into the atmosphere. Because of the low sound speed in this gas (about 100 m/s), the flow was internally supersonic. Old Faithful Geyser, Wyoming, is a converging-diverging nozzle in which liquid water refilling the conduit during the recharge cycle changes during eruption into a two-phase liquid-vapor mixture with a very low sound velocity. The high sound speed of liquid water determines the characteristics of harmonic tremor observed at the geyser during the recharge interval, whereas the low sound speed of the liquid-vapor mixture influences the fluid-flow characteristics of the eruption. At the rapids of the Colorado River in the Grand Canyon, Arizona, supercritical flow occurs where debris discharged from tributary canyons constricts the channel into the shape of a converging-diverging nozzle. The geometry of the channel in these regions can be used to interpret the flood history of the Colorado River over the past 10^3-10^5 years. The unity of fluid mechanics in these three natural phenomena is provided by the well-known analogy between gas flow and shallow-water flow in converging-diverging nozzles.

I. Introduction: geologic nozzles

An eruption of Old Faithful geyser, a flood on the Colorado River, and a lateral blast from Mount St. Helens do not, at first glance, appear to be related. A geographic map of the locations of these three places certainly does not reveal any underlying geologic unity (Fig. 1). However, a fluid-dynamical unity is revealed when the "locations" are shown instead on a schematic diagram[2] of gas flowing through a nozzle or shallow water flowing through a flume (Fig. 2). The analogy between the flow fields for compressible gas and shallow water is semiquantitative and was thus widely explored in the early days of wind-tunnel development[3]. In modern times the analogy has been

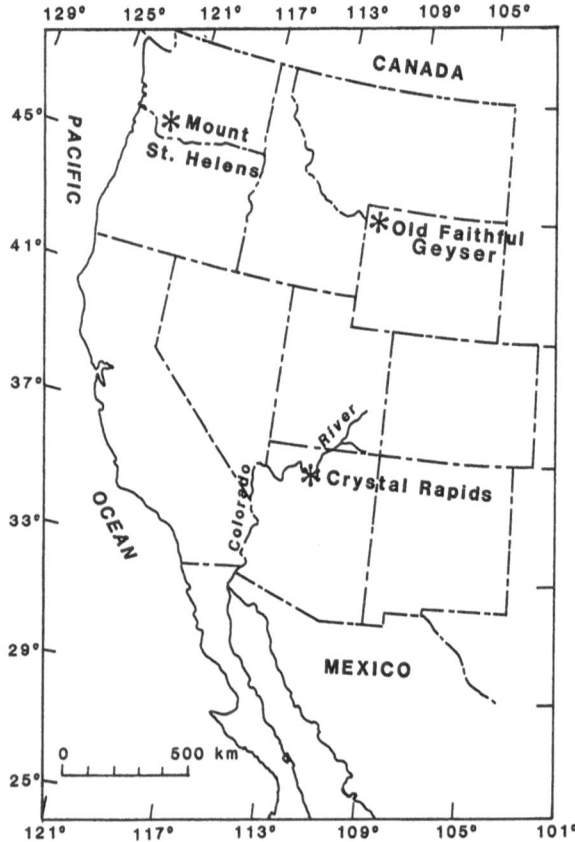

FIG. 1. Index map of the geographic locations of Crystal Rapids (Grand Canyon, Arizona), Old Faithful Geyser (Yellowstone National Park, Wyoming), and Mount St. Helens (Washington).

primarily a teaching tool[4] and has never been used by geologists to explain large-scale natural phenomena. The purpose of this paper is to show the basis for invoking nozzle-flow theory for interpretation of complex geologic events and to provide a perspective on geological problems in which the importance of supercritical and supersonic flow has been underestimated.

A major reason that geologic events have not been viewed from the particular perspective of fluid mechanics presented here is the subdivision of fluid mechanics and its applied fields into the specialties of compressible and incompressible flow; for example, aeronautics versus hydraulics. This subdivision arises from the need to simplify the complex momentum and continuity equations in order to solve practical problems. The momentum equation for a viscous fluid moving in a gravitational field under the influence of a pressure gradient is complex because of dimensionality and nonlinearity:

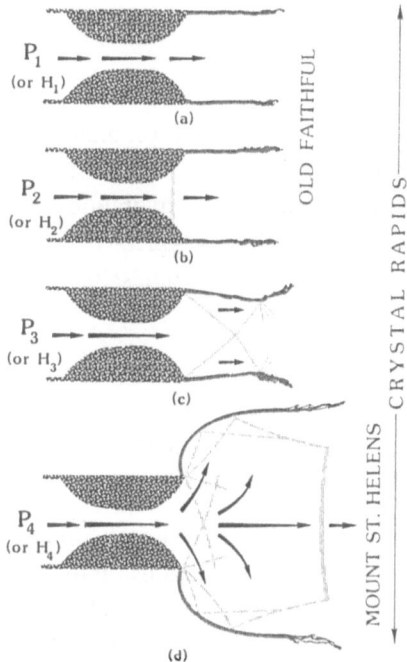

FIG. 2. Four diagrams showing the behavior of gas flowing through a converging-diverging nozzle, or of shallow water flowing through a converging-diverging flume; arranged vertically. The nozzle walls are shown by the heavy lines and heavy shading at the left; the flow direction is indicated by arrows, and the velocity magnitude is indicated schematically by the length of the arrows. The reservoir of gas or liquid is on the left: $P_1, P_2, \ldots$ represent increasingly larger reservoir pressures, and $H_1, H_2, \ldots$ represent increasingly larger values of hydraulic head, compared to ambient or downstream conditions on the right. At the exit of the nozzle (or channel) on the right, the structure of the flow field in the departing fluid is shown schematically by medium shading. Shock and rarefaction waves (alternatively, positive and negative normal and oblique hydraulic jumps) are indicated by the lightest shading. Relative flow conditions are shown schematically by the position of the lettering at the right: Old Faithful is subsonic (a) or weakly supersonic (b), Mount St. Helens is strongly supersonic (d), and Crystal Rapids involves all conditions (a-d) from subcritical to strongly supercritical.

$$\rho \frac{D\mathbf{u}}{Dt} = -\nabla P + \nabla \cdot \tau + \rho \mathbf{g} \ . \tag{1}$$

In this equation, ρ is the fluid density, D/Dt is the material derivative, $\mathbf{u}$ is the fluid velocity, ∇P is the pressure force acting on the fluid, $\nabla \cdot \tau$ is the viscous force, and $\mathbf{g}$ is the acceleration of gravity. The continuity equation for mass is generally simpler, but is still difficult to apply in a geometrically complicated problem:

$$\frac{D\rho}{Dt} = -\rho \left(\nabla \cdot \mathbf{u} \right) .$$

(2)

In many cases these two important equations can be considerably simplified by consideration of the fluid properties or the boundary conditions of the problem. For example, if pressure changes are relatively small, compressibility can be neglected, so that $\rho \sim$ constant and $\nabla \cdot \mathbf{u} = 0$. Such an assumption underlies all of hydraulics, and geologists with interests in hydraulics or related geomorphic problems typically diverge at an early stage of their education from advanced studies of compressible fluid dynamics.

Alternatively, in many flows the pressure gradient may be great enough so that compressibility is important but gravity is not; $\mathbf{g} \sim 0$. This latter condition is assumed in most of gas dynamics and, because of the prominent role of gravity in most geologic processes, few geologists are exposed to a rigorous gas-dynamics curriculum.

Although the subjects of nozzle gas dynamics and of shallow-water hydraulics evolve from very different approximations to the conservation equations, important concepts common to both subjects have been recognized because, when reduced to suitable nondimensional variables, the conservation equations in the two subjects become identical. (Readers familiar with this identity can skip directly to Section II.)

Examine first the mass and momentum equations for a perfect gas. For simplicity, assume that the flow is quasi-one-dimensional along a coordinate direction x. The equations of mass and momentum conservation for flow of a compressible gas are

$$\frac{\partial \rho}{\partial t} + u \frac{\partial \rho}{\partial x} + \rho \frac{\partial u}{\partial x} = 0$$

(3)

and

$$\frac{\partial u}{\partial t} + u \frac{\partial u}{\partial x} + \frac{1}{\rho} \frac{\partial P}{\partial x} = 0 .$$

(4)

For a perfect gas and isentropic flow,

$$P V = R T$$

(5)

and

$$P V^{\gamma} = P_o V_o^{\gamma} = \text{constant} ,$$

(6)

where P is pressure, V is volume, R is the gas constant, T is temperature, and γ is the ratio of specific heats (the isentropic exponent). The subscript o indicates a reference state (typically one where the fluid is at rest with velocity $u = u_o = 0$). For a perfect gas, Eqs. (3)-(6) can be combined to give

$$\frac{\partial u}{\partial t} + u \frac{\partial u}{\partial x} + \frac{\gamma P_o}{\rho_o^\gamma} \rho^{\gamma-2} \frac{\partial \rho}{\partial x} = 0 \,. \tag{7}$$

For water flowing from one infinite reservoir into another with lower head, the equations of motion that can be directly compared with Eqs. (3) and (7) are

$$\frac{\partial h}{\partial t} + u \frac{\partial h}{\partial x} + h \frac{\partial u}{\partial x} = 0 \tag{8}$$

and

$$\frac{\partial u}{\partial t} + u \frac{\partial u}{\partial x} + g \frac{\partial h}{\partial x} = 0 \,, \tag{9}$$

where h is the water depth. In these equations, and in the figures in this paper, it is assumed that in a vertical cross section (containing the coordinate z), the bottom of the water is at the channel boundary, $z = 0$, and the water has a free surface at $z = h$. The free surface is assumed to be at constant atmospheric pressure, P_a, but its elevation can vary along the channel and with time. The velocity of the water can also vary with position along the channel, but is assumed constant over any vertical cross section. The viscosity and compressibility of the water are ignored. Again, as with gas flow, quasi-one-dimensional flow is assumed -- the contours of the channel walls must be gradual, and vertical accelerations of the water must be small compared with the acceleration of gravity, g.

The shallow-water conservation equations (8) and (9) are identical to the compressible-gas conservation equations (3) and (7) *if* (1) water depth, h, is analogous to gas density, ρ, and (2) the isentropic exponent, γ, of the gas is equal to 2 (a value which, unfortunately, is never attained in real gases, so that the mathematically equivalent flow fields cannot be physically realized). Examination of the conservation equations and the equations of state for a perfect gas shows that h is also analogous to T, and h^2 is analogous to P. Hence, within the context of all of the simplifying assumptions, flow of gas in a nozzle and flow of shallow water in a flume are governed by the same conservation equations. Identical flow fields therefore occur when the proper nondimensional variables are considered.

The illustrations in Fig. 2 represent the flow conditions at different ratios of upstream (reservoir) and downstream (atmosphere or tailwater) conditions. If the figure is interpreted as representing a cross section of a horizontal nozzle through which gas is flowing from left to right, the different parts of the figure represent the flow field from different high pressures $(P_1, P_2, \ldots)$ in the left reservoir. The gas flows into an infinitely large reservoir (not shown) at lower pressure on the right. Alternatively, if Fig. 2 is interpreted as representing a map view of a horizontal channel in which shallow water is flowing from left to right, the different parts of the figure represent the flow field from reservoirs of different depth. The driving energy for the flow is the elevated depth of water in the left reservoir compared with the right. The water has a potential energy, H_r, called the *head* and generally expressed as a depth, indicated as $H_1, H_2, \ldots$ in the parts of Fig. 2.

With this introduction, let us reexamine the sense in which each of the geologic problems mentioned above is a nozzle problem[5]. The nozzle of the Colorado River is the river channel, a converging-diverging nozzle formed by debris flows that constrict the main channel, and the fluid is shallow water. The "geologic twist" that complicates simple application of flume concepts is that the walls and bed of the channel are erodible, and the channel can therefore change shape in response to changing conditions in the flow. The nozzle of Old Faithful geyser is a fissure of irregular (and largely unknown) geometry extending more than 20 m into the ground. The geologic twist in this problem is that the fluid is much more complex than a perfect gas: hot, liquid water stands in the conduit between eruptions, and then boils and changes through a complex unloading process into a droplet-laden steamy aerosol during an eruption. The nozzle of the Mount St. Helens lateral blast was a huge vent created when a landslide caused by an earthquake opened a vertical scarp nearly 0.25 km^2 in area and exposed a hot, hydrothermal, magmatic system. The erupting fluid was a hot vapor heavily laden with ash, rocks, ice fragments, and tree debris. As these three examples show, the scale of the geologic nozzles is large, the nozzle shapes are irregular, and the thermodynamic properties of the flowing fluids are complex.

II. Sound velocities and critical velocities: their influence on the flow field

The most important result from the above analogy is the recognition that characteristic velocities control flow behavior in shallow water and gas flow. For small disturbances, the equations of momentum (Eqs. (7) and (9)) can be linearized and written as

$$\frac{\partial^2 P}{\partial t^2} - a_o^2 \frac{\partial^2 P}{\partial x^2} = 0 \qquad \text{(perfect gas)} \qquad (10)$$

and

$$\frac{\partial^2 h}{\partial t^2} - g\, h_o \frac{\partial^2 h}{\partial x^2} = 0 \qquad \text{(shallow water)} . \qquad (11)$$

These are the well-known *wave equations*, from which it can immediately be seen that small disturbances propagate with characteristic velocities proportional to the square root of the coefficient of the second term. In compressible gas flow, the characteristic velocity is the *sound velocity*, a_o, the velocity at which small perturbations in density or pressure propagate through the fluid:

$$a^2 = \left[\frac{\partial P}{\partial \rho} \right]_S , \qquad (12)$$

where the derivative is taken at constant entropy, S. In shallow-water flow, the characteristic speed is the *critical velocity* -- the velocity of a gravity wave of long wavelength and infinitesimal strength:

$$c^2 = g\,h . \qquad (13)$$

In both cases, the nature of the flow field depends on the magnitude of the fluid velocity compared with the characteristic velocity. The *Mach number*, M, of a compressible gas flow is the ratio of the mean flow velocity to the sound speed:

$$M = u/a . \qquad (14)$$

The *Froude number* of shallow water flow, Fr, is the ratio of the mean flow velocity to the critical velocity:

$$Fr = u/c . \qquad (15)$$

The local flow variables are determined by these dimensionless ratios, which, in turn, depend on reservoir conditions and geometry. For gas flow, the important parameters are the ratio of the pressures in the driving and receiving reservoirs, the area ratio along the axis, and the gas equations of state (particularly R and γ for a perfect gas). For shallow-water flow, the important parameters are the ratio of upstream to downstream energy and the area ratio of the channel. Depending on the values of these parameters, the flow field can have dramatically different properties, as illustrated in Fig. 2.

Consider, first, that Fig. 2 represents the flow of gas through a nozzle. When the pressure P_1 in the reservoir is "low"[6], the fluid accelerates from the reservoir into the constriction and decelerates in the diverging section (Fig. 2a). This is the classic *venturi tube*, and the flow is everywhere *subsonic*.

If the pressure ratio is higher (Fig. 2b), the fluid accelerates from the reservoir into the converging section and can reach *sonic* or *choked* conditions $(M = 1)$ in the throat; it can be rigorously shown that sonic conditions can only occur in the throat. At one particular pressure ratio the flow can decelerate back to subsonic conditions in the diverging section, but for higher values it will accelerate to *supersonic* conditions in the diverging section. Strong nonlinear waves -- *shock* and *rarefaction waves* -- can be present and are, in fact, usually required to decelerate the flow back to ambient conditions in the exit reservoir. At pressure ratios for which supersonic flow conditions are obtained, a *normal shock* stands in the diverging section and the deceleration to ambient conditions occurs within the nozzle between the shock and the exit plane (Fig. 2b). At still higher pressure ratios (Figs. 2c and 2d) the shock is "blown out" of the nozzle, and a complicated flow field consisting of oblique and normal shocks and mixed regions of subsonic and supersonic flow exists within the exiting jet. Because the decelerating waves are nonlinear, the jet "overshoots" ambient conditions and multiple shock and rarefaction waves are required to achieve the pressure balance[7].

Consider alternatively that Fig. 2 represents shallow-water flow. When the head difference between the reservoirs on the left and right is "small"[8], the flow is *subcritical* everywhere -- the fluid accelerates in the converging section and through the constriction, and decelerates in the diverging section (Fig. 2a). The flow field is analogous to that in a venturi tube. The word *streaming* is often used for subcritical flow.

If the head ratio is greater, as indicated in Fig. 2b, the flow accelerates from the reservoir through the converging section and can reach *critical* conditions ($Fr = 1$) in the constriction. At the critical value of head ratio the flow can decelerate to subcritical conditions in the diverging section, but for other higher values it will accelerate to *supercritical* conditions in the diverging section. The word *shooting* is often used for supercritical flow. Strong nonlinear waves, in this case called *oblique* (or *slanting*) and *normal hydraulic jumps*, are generally required to decelerate the flow back to ambient conditions in the downstream reservoir. Depending on the head ratio and the severity of the constriction, waves can stand in (Fig. 2b) or downstream of (Figs. 2c and 2d) the diverging section[9].

The flow fields shown in Fig. 2 are a subset of possible flow conditions, for they do not show possible wave structures that arise if fluid enters the constricted part of the nozzle in a supersonic or supercritical state. Such conditions can, in fact, be obtained geologically. For example, if a change in river-bed elevation causes water to accelerate to supercritical conditions before a lateral constriction is encountered, the flow can be supercritical as it enters the convergence, rather than subcritical as illustrated in Fig. 2. For simplicity, this complexity is ignored in this paper.

Supersonic or supercritical conditions are amazingly easy to obtain in geologic settings. If the ratio of reservoir pressure to atmospheric pressure in a gas nozzle is more than about 2, sonic and supersonic flow will occur in the nozzle; for comparison, the ratio of pressure in a volcanic reservoir to atmospheric pressure is often around $100:1$. If shallow water flows from one reservoir to another that has less than 2/3 of the head of the source reservoir, critical conditions can be obtained in the throat; for comparison, backwater depths on the Colorado River may exceed downstream tailwater depths by a factor of 2. Thus, the existence of supersonic or supercritical flow fields in geologic settings is conceptually reasonable.

Our intuition, however, generally fails to prepare us for the possibility of such flows in the natural world. We commonly think of supersonic flow in terms of modern aeronautics: objects obtain high Mach numbers by moving very fast through air, which has a high sound speed. Geologic fluids rarely move at the speeds characteristic of modern aircraft (except in some volcanic eruptions), but the entire spectrum of flow behavior from subsonic to supersonic (and subcritical to supercritical) can occur in geologic flows because the fluids can have very low characteristic velocities. Fluids with low sound velocities can develop internally supersonic flow fields while still moving subsonically with respect to the surrounding atmosphere. That is, there can be standing shock or rarefaction waves internal to the flow, but no standing waves in the external medium.

Fluids in geothermal and volcanic settings typically have low sound speeds: water that contains gas bubbles (e.g., air or CO_2) or steam bubbles (boiling water) has a very low sound speed, because the gas bubbles dramatically increase the compressibility of the mixture, κ_S. An alternative form of the definition of sound speed, $a = (1/\kappa_S)^{1/2}$, shows this dependence clearly. The sound speed in an air-water mixture can be as low as 20 m/s. The sound speed is further decreased in a mixture in which the bubbles are of the same composition as the liquid (e.g., steam bubbles in boiling water), because exchanges of mass and latent heat accompany passage of a sound wave; these exchange processes also decrease the sound velocity. Sound speeds as low as 1 m/s are possible for boiling water[10].

The dependence of sound speed on phase, and on pressure and temperature, can be shown on an entropy-density (S-ρ) phase diagram[11] (Fig. 3). This representation is suggested by the definition of sound speed given in Eq. (12) above: on a graph of density versus entropy, sound speed is proportional to the vertical gradient of isobars. Such a graph can be read as an ordinary topographic map on which "flatlands" represent low sound speeds and "cliffs" represent high sound speeds (shown schematically in the inset in Fig. 3). The S-ρ representation shows the wide range of sound speeds characteristic of simple one-component substances. If such fluids flow from high to low pressure (e.g., in eruptions or in geo-thermal wells), the phase of the fluid can change from liquid to liquid + vapor, or from vapor to vapor + liquid. A hypothetical decompression path appropriate to Old Faithful (and discussed later in Section IV) is shown as the vertical line (a) in Fig. 3. Note that along this decompression path the sound speed can change by several orders of magnitude. If the fluid is in a two-phase state, flow velocities of only a few tens of meters per second can give a wide range of Mach numbers, including sonic ($M = 1$) and supersonic ($M > 1$) flow.

Mass loading of vapor with solids or liquid droplets also produces fluids with low sound speeds. No data or theories exist for the fluids encountered in volcanic problems, where, for example, the mass ratio of solids to vapor can exceed 100, and particle sizes within a single flow can range from microns to meters. At present, we can only apply simple pseudogas theory to this problem to obtain characteristic sound speeds (Fig. 4). Sound speeds of 50 to 100 m/s are plausible. Flow velocities in gassy volcanic eruptions are commonly on the order of 100 m/s and can exceed 500 m/s. Therefore, a wide range of Mach numbers, including $M > 1$, can be obtained.

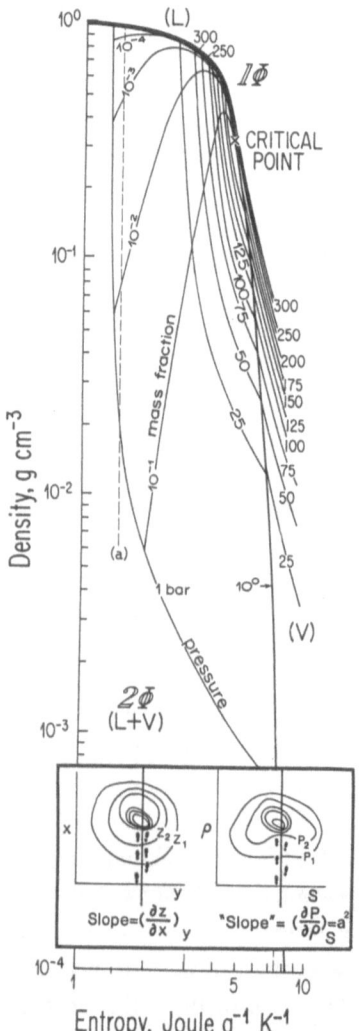

FIG. 3. The entropy-density $(S - \rho)$ phase diagram[11] for H_2O. Entropy is relative to the triple point of H_2O. The label 1ϕ marks the single-phase field and the label 2ϕ marks the two-phase field (liquid + vapor). Contours of constant pressure (isobars) are shown in 25-bar increments. In the two-phase field, contours of constant mass fraction of vapor (isopleths) are shown. A similar graph for CO_2 can be found in Ref. 11. The vertical dashed line (a) represents an isentropic path that might be taken by a fluid decompressing from the bottom of Old Faithful. The horizontal inset is a comparison of a standard topographic map of a conical hill (left) with the entropy-density graph used to illustrate sound speed. The spacing of the contours is inversely proportional to the steepness of the topography in the case of a map (left), and to the sound speed in the case of an $S - \rho$ diagram (right). In the latter case, the "footsteps" indicate that the appropriate derivative for considering sound speed is in the vertical direction.

Finally, note that the critical velocity in shallow-water flow plays the same role as the sound speed in determining transitions between linear and nonlinear flow regimes. Critical velocities in rivers can be of the order of the flow velocities, even in major rivers where large depths increase the critical velocity (Eq. 13). In the Colorado River in the Grand Canyon, for example, water depths of the order of 10 m are common; the corresponding critical velocity is 10 m/s. In most calm stretches of the river, flow velocities are on the order of 1 m/s, and Froude numbers are less than 0.1. In major rapids, however, where the water becomes shallow and fast, the flow velocity can exceed the critical velocity $(Fr > 1)$. Supercritical flow is not common in rivers[12], but when it does occur, the geologic consequences can be great; one such case is discussed in the following section.

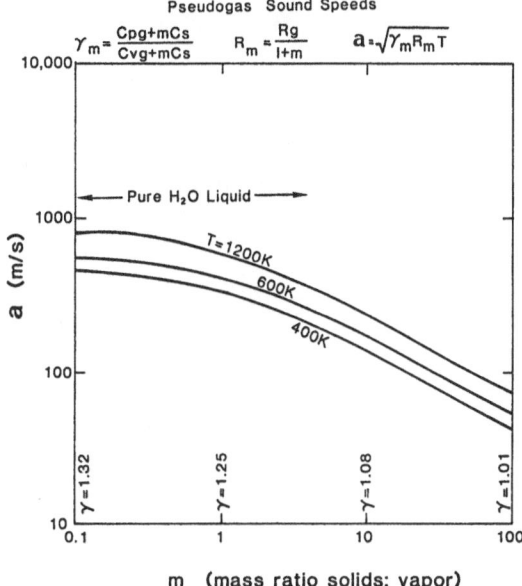

FIG. 4. Sound speed, a, of pseudogas versus mass ratio, m, of solids to vapor (steam). Curves for three different temperatures spanning the range of geothermal and volcanic interest are shown. The sound speed of pure liquid water is indicated. The isentropic exponent, γ, varies with m as indicated at the bottom of the graph. Because the mass loading is high (of order 10, say) in many volcanic eruptions, γ is near unity.

Many simplifications have been made in the discussion above, and these, as well as others, will be used in the analyses below: e.g., thermodynamic equilibrium; isentropic, quasi-one-dimensional flow; steady flow; and perfect gas or pseudogas behavior. One additional major simplification in the following analyses is that the flow fields are assumed to be *either* compressible and gravity-free ($M > 1$, $Fr > 1$), or gravity-dominated and incompressible ($Fr < 1$, $M < 1$). The assumption of incompressible shallow-water flow is good for the Colorado River. However, compressibility and gravity are probably both important for the flow fields of Old Faithful and the Mount St. Helens lateral blast (i.e., $M > 1$, $Fr < 1$)[13]. This complex problem is only beginning to be addressed as the capabilities of modern supercomputers are being turned toward the problem.

III. Crystal Rapids: subcritical and supercritical flow in an erodible channel

A. Geologic setting and the events of 1983

The Colorado River is the largest of the great rivers in western America. In the 400-km stretch through the Grand Canyon, numerous debris fans have been deposited

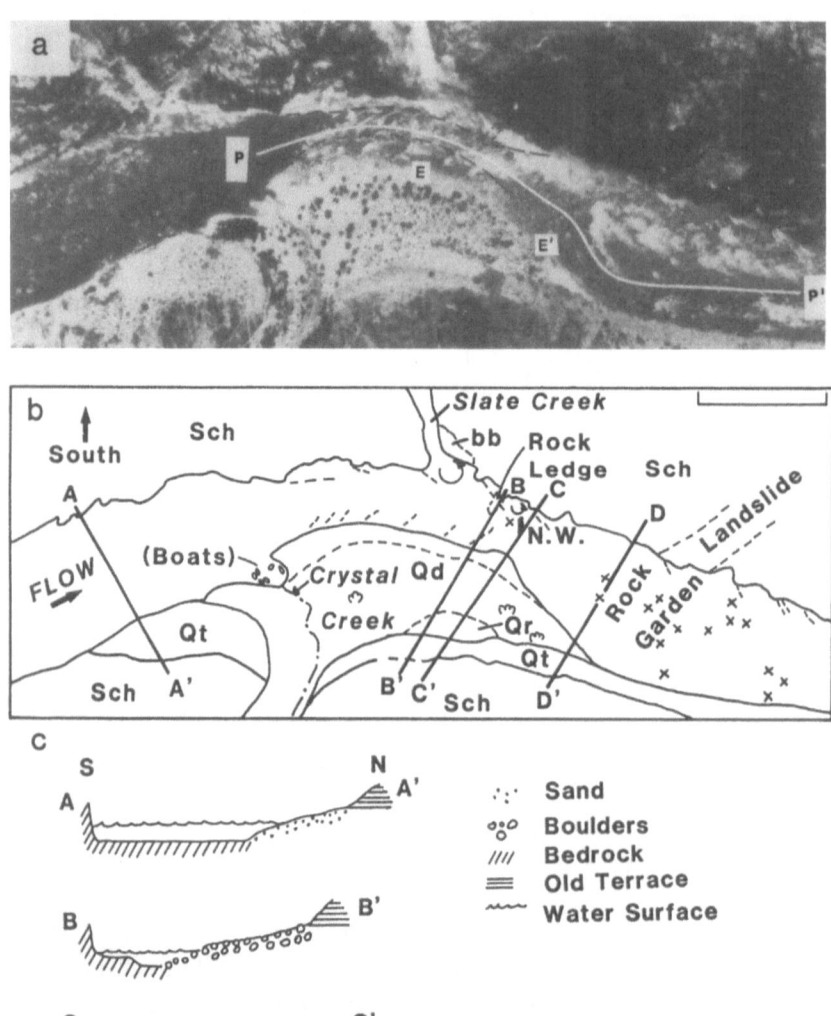

FIG. 5. (a) and (b) Crystal Rapids at two dramatically different discharges. (c) and (d) Keys to features in (a) and (b). (a) June 16, 1973 (U.S. Geological Survey Water Resources Division air photo). The discharge in (a) was about 283 m³/s (10,000 cfs); the discharge in (b) was 2,600 m³/s (92,000 cfs). The rise of the river shoreline from the lower (southern) end of the debris fan in (a) to the base of the old alluvial terrace in (b) represents a stage change of about 5.5 m. Note the constriction of the river channel as it passes downstream from left to right through the debris fan, and the expansion downstream of the debris fan. In the downstream region (beginning approximately at the inflection point on the right half of PP′ and extending past P′ in (a)), the channel

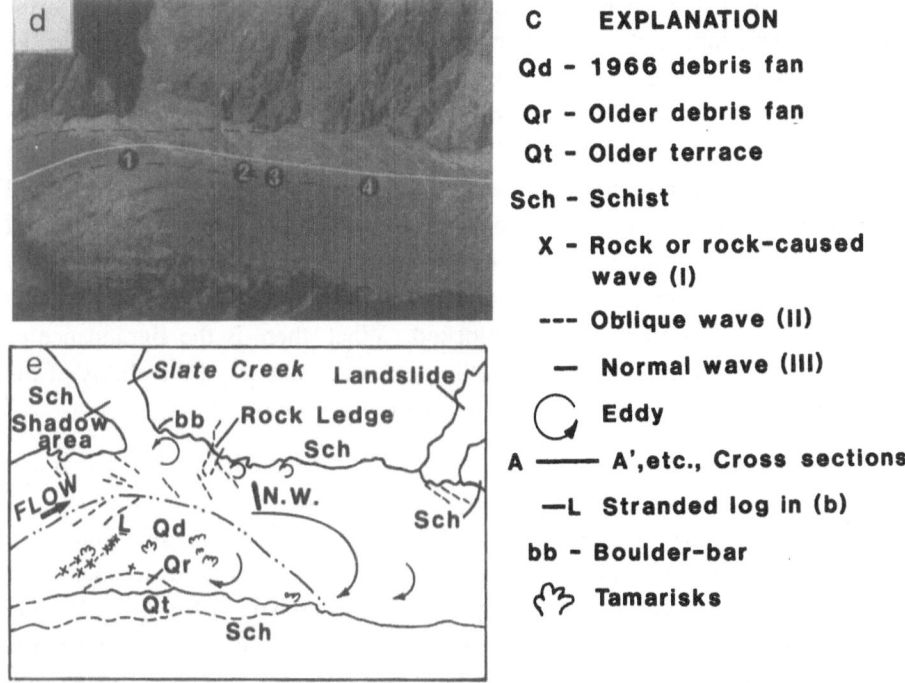

C EXPLANATION

Qd - 1966 debris fan

Qr - Older debris fan

Qt - Older terrace

Sch - Schist

X - Rock or rock-caused wave (I)

--- Oblique wave (II)

— Normal wave (III)

Eddy

A ——— A',etc., Cross sections

—L Stranded log in (b)

bb - Boulder-bar

Tamarisks

FIG. 5. (continued)

bottom is littered with boulders (the "rock garden" whose origin is discussed in the text). Rocks in the rock garden are visible at 283 m^3/s (10,000 cfs) as shown in (a), cause substantial waves at 30,000 cfs, and are submerged at 2600 m^3/s (92,000 cfs) as shown in (b). PP' was the preferred navigation route prior to 1983. The normal hydraulic jump of interest in this paper (indicated by NW in (c) and (d)) is not easily visible in the two photographs, both because of the large area covered by the photos and because turbulence and many small waves cause variations in the reflectivity of the features on the water. The wave can be seen in detail in Fig. 7. (c) Schematic cross-sections. The relative widths are correct; the vertical scale is arbitrarily exaggerated. The important points to note from the cross sections are the small cross-sectional areas at BB' caused by high velocities in the constricted part of the rapid, and the greater depth in CC' caused by the hydraulic jump (shown schematically as an exaggerated wave on the water surface).

by flash floods in tributary canyons (Fig. 5). Flash floods in the tributaries can carry boulders many meters in diameter into the path of the Colorado River, because the gradient of the tributaries is quite steep. When emplaced, the large debris fans temporarily obstruct the path of the river, damming it until the debris deposit is breached and a new channel carved. The major rapids on the Colorado River are located where the river passes through these debris fans.

The channel of the Colorado River resembles a converging-diverging nozzle in the vicinity of these debris fans (note the constriction of the channel in Fig. 5). Typically, the channel narrows from a characteristic upstream width of about 100 m to a narrowest point in the "throat" of the rapid, and then diverges back to a downstream width about equal to the upstream width (e.g., Fig. 5). The ratio of the width of the river at the throat to the width at an average upstream section is the *constriction* of the river; I will also refer to it as a "shape parameter." Constrictions at the debris fans in the Grand Canyon are remarkably uniform at a value of about 0.5 (Fig. 6). There is no *a priori* reason to believe that the debris fans themselves were emplaced in such a way that, by coincidence, half of the main channel was blocked. What, then, is the significance of this characteristic nozzle shape? It must be telling us something about the ability of the Colorado River to erode its own channel, i.e., to contour its own nozzle.

Because the debris fans are generally very old (of the order of 10^3 to 10^5 years) and because the flash floods that create and renew them are rare, we have little hope of observing the processes that create the balances between tributary floods and main channel erosion. However, a unique series of events spanning the two decades from 1966 to 1986 has given us a glimpse of these processes. In January, 1963, Glen Canyon Dam at Page, Arizona (Fig. 7) was closed, and the discharge into the Colorado River through the Grand Canyon became controlled by demands for electrical power and water storage at the dam, rather than by natural flooding. About four years after the dam was closed, a flash flood down Crystal Creek emplaced a large debris fan across the river about 175 km below the dam (Fig. 5). There were no witnesses to this event;

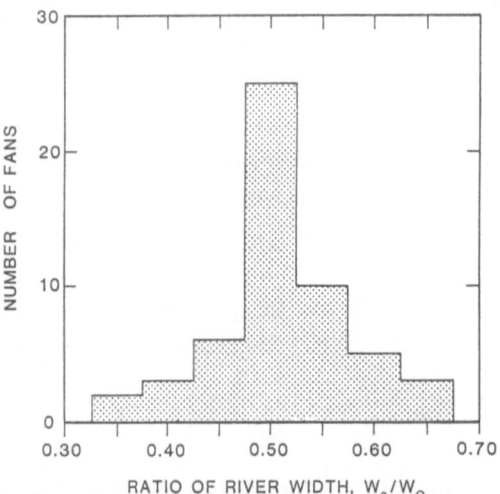

FIG. 6. Histogram of constriction values (shape parameter) for the Colorado River as it passes 54 of the largest debris fans in the 400-km stretch below Lee's Ferry, Utah. These values are based on the widths of the *surface* water in the channel in 1973 air photos (such as shown in Fig. 5).

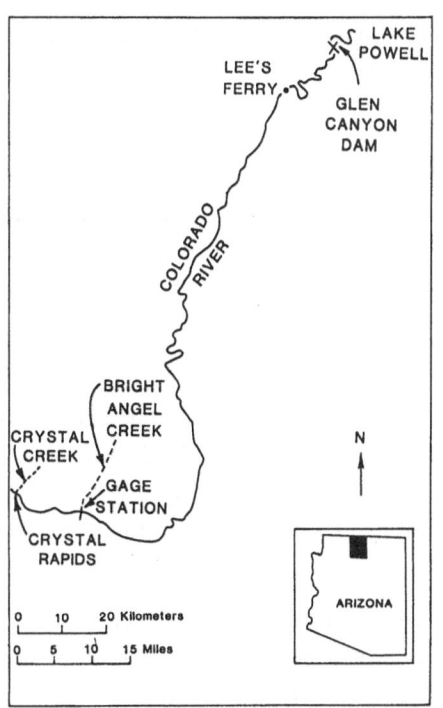

FIG. 7. Index map for locations near Crystal Rapids.

by the time observations were made, the Colorado River had carved a channel through the distal (south) end of the debris fan. When first measured (on the 1973 air photograph shown in Fig. 5), the constriction of the river channel through the fan[14] was roughly 0.33, substantially more severe than the constrictions of the more mature fans along the Colorado River. From 1966 to 1983, the discharges through the dam were held below 850 m³/s (30,000 cubic feet per second (cfs))[15], and the constriction remained at about 0.33.

The water surface of the Colorado River became very rough and turbulent as it passed through the Crystal debris fan -- this stretch of water, nearly 1 km long, is known as Crystal Rapids. The boulders, waves, and eddies in Crystal Rapids made it one of the two most difficult stretches of the river for raft navigation, even at the normal levels of controlled discharges between 1966 and 1983 (140 m³/s (5,000 cfs) to 850 m³/s (30,000 cfs)). The rapid is a major hazard for recreational rafting, an activity in the Grand Canyon involving about 10,000 people each year. Between 1966 and 1983, the major navigational obstacle occurred where water poured over a large rock into a deep hole and emerged through a sharp-crested wave in the narrowest part of the rapid. This feature was known as the "Crystal Hole" (the location of the Crystal Hole is shown in Fig. 5, but the feature itself is too small to show at the scale of this photograph).

At low discharges, a rock about 2 m high was seen as the cause of this hydraulic feature, and the significance of this rock relative to the significance of the severe constriction was, in hindsight, overestimated. Many waves in the rapids of the Colorado River are caused by large rocks. River guides who ran the river before Glen Canyon Dam was closed, when large natural floods reached 2,300 to 3,600 m³/s (80,000 to 125,000 cfs) annually, reported that, in most rapids, the waves became very weak or disappeared ("washed out") at high discharges because the obstacles causing them became submerged ("drowned"). There was, however, no record of the behavior of waves in Crystal Rapids at discharges exceeding 850 m³/s (30,000 cfs), because the rapid -- in its modern severe form -- did not exist before construction of Glen Canyon Dam.

In 1983, rapid snow melt in the headwaters of the Colorado River forced the Bureau of Reclamation to increase discharges through Glen Canyon Dam to 2,600 m³/s (92,000 cfs) to prevent Lake Powell from flowing over the dam. As discharges increased above 850 m³/s (30,000 cfs) -- a level that had not been exceeded for two decades -- the waves in most rapids disappeared, as expected. The rapids "drowned out" and the river ran smooth and fast through most of the Grand Canyon.

This was not the case at Crystal Rapids: as the discharge reached 1,700 to 2,000 m³/s (60,000 to 70,000 cfs), a wave reported by experienced boatmen to have been as high as 9 m, and photographically documented to have exceeded 5 m, stood across most of the river channel (Fig. 8). At greater discharges the height diminished -- at 2,600 m³/s (92,000 cfs) the wave surged only between 3 and 5 m. Because typical river rafts are 5 to 11 m in length and 2 to 4 m in width, the wave was a severe obstacle to boating. One rafter was drowned, and dozens of others were seriously injured. The National Park Service closed the rapid to commercial boating until the discharges were decreased. The existence of this large wave at high discharges, and its evolution with changing discharge, provided clues about the relation between the Colorado River and its debris fans.

B. Results of analysis of shallow-water flow

Certain aspects of the flow in the rapids of the Colorado River can be analyzed in terms of conservation of mass and momentum for flow in a converging-diverging channel[16]. As discussed in the context of Fig. 2, the shape of the channel, and the upstream and downstream reservoir heads, must be specified. The river channel, which is actually very irregular in shape (Fig. 5c), is simplified to be rectangular in cross-section for the analysis. This simplification causes the shape parameter (the average constriction) to be 0.25, instead of the value 0.33 measured from the surface width of the water.

Six different flow zones can be identified, as shown in Fig. 9a: (0) an upstream state of unconstricted, uniform flow; (1) the convergent section of the channel upstream from the constriction; (2) the constriction; (3) the beginning of the divergence; (4) the end of the divergence; and (5) a downstream state of uniform flow not influenced by the constriction. Regions (3) and (4) may be separated by a hydraulic jump.

FIG. 8. River raft (a) entering and (b) trapped in the large wave at Crystal Rapids on June 25, 1983, when the discharge was approximately 1,700 m³/s (60,000 cfs). Photographs copyrighted by Richard Kocim; reprinted with permission. The pontoons on the raft are each 1 m diameter; the midsection is about 3 m diameter. More than 30 passengers are on board; one head is visible on the lower left side of the raft in (b). To aid the reader in distinguishing the water surface in the foreground and near the boat from turbulent water in the background, a white line has been drawn along the approximate surface of the water upstream from and through the hydraulic jump in (a). From the scale of the raft, the trough-to-crest height of the wave can be estimated to exceed 5-6 m.

Water flows through the rapid because the upstream reservoir is higher than the downstream reservoir. In a notation slightly changed from that used in the general equations of Section I, to be consistent with Fig. 9b, application of the Bernoulli equation to any two cross sections 1 and 2 gives the total energy balance as

$$z_1 + D_1 + \frac{u_1^2}{2g} = z_2 + D_2 + \frac{u_2^2}{2g} + E , \qquad (16)$$

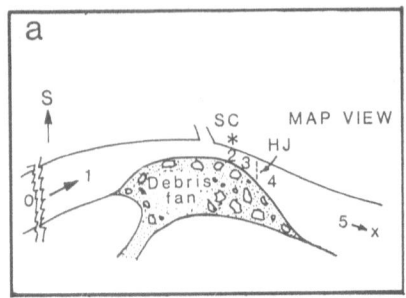

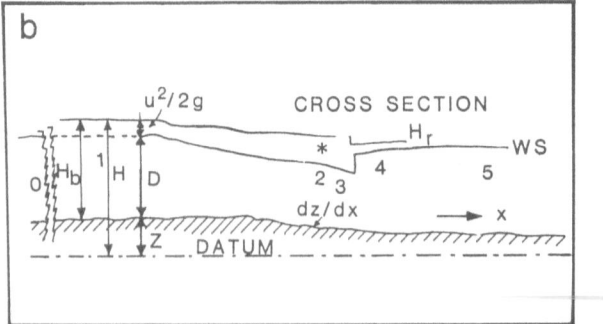

FIG. 9. (a) Schematic map view of the river and the debris-fan configuration at Crystal Rapids. The debris fan at the bottom of the map emanates from Crystal Creek. SC indicates the position of the other tributary, Slate Creek. The regions 0, 1, 2, 3, 4, 5 are defined in the text. HJ indicates a possible hydraulic jump. (b) Schematic longitudinal profile, showing the notation used in the energy relation given in the text and in Fig. 10.

where E is the energy dissipated between sections 1 and 2. (The change in convention from the variable h used in Section I for water depth to the variable D is required to account for a sloping channel bottom by referencing both the channel bottom and the water surface to a datum plane, as in Fig. 9b). For simplicity, and because of a paucity of data, the energy change due to the change in bed elevation $(z_1 - z_2)$ is assumed to be compensated by energy dissipation in the flow, E. This assumption allows the flow to be considered at constant *specific energy* $(D + u^2/2g)$, except across hydraulic jumps. For the analysis, the discharge variation of the specific energy of the unconstricted flow upstream of Crystal Rapids was estimated from measurements at a U.S. Geological Survey gage station 16 km upstream at Bright Angel Creek (Fig. 10a, heavy line, H_r). The flow below Crystal Rapids was assumed to return to this same specific head.

The flow entering the rapid can have the ambient specific head, H_r, if and only if all of the discharge can be accommodated through the constriction. If the constriction is

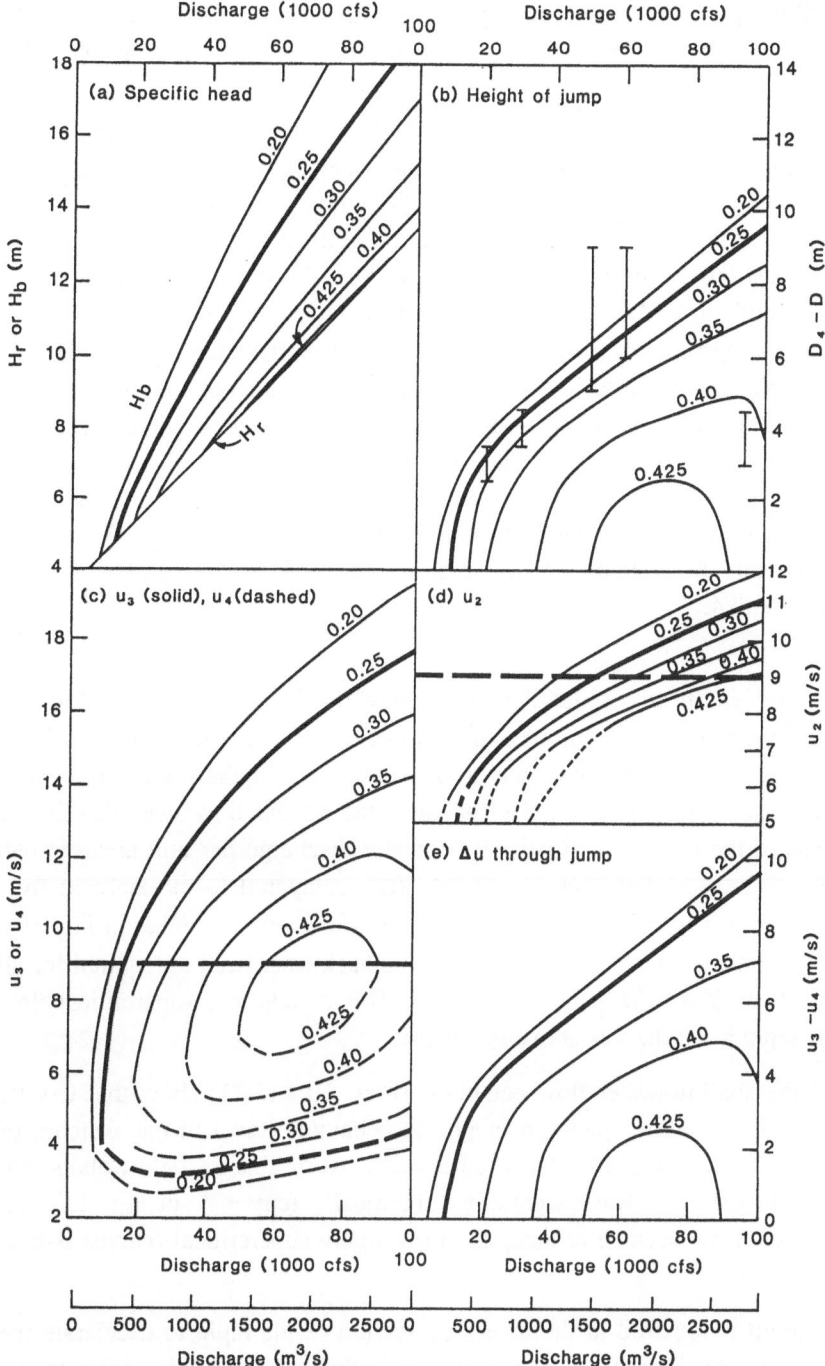

FIG. 10. Summary of the shallow-water flow calculations for Crystal Rapids given in Kieffer[16]. Subscripts on the various parts of this figure refer to regions defined in the text and shown schematically in Fig. 9; e.g., u_2 is the flow velocity in flow zone 2. In

all sections of this figure, the curve appropriate to the initial constriction at Crystal Rapids, 0.25, is shown by a heavy line. (a) Specific head, H_r, of the unconstricted river versus discharge. The curves labeled H_b are the backwater heads that develop upstream of Crystal for conditions of critical (choked) flow. The backwater head depends on the constriction and is given for values ranging from 0.20 to 0.425, as labeled on the curves in this and all other parts of the figure. (b) The curves show the calculated height of the hydraulic jump that separates regions 3 and 4 when supercritical flow occurs; the constrictions are indicated on each curve. The bars are values observed during the 1983 flood. (c) The solid curves (top) are calculated values of the flow velocity in the diverging section of the channel (region 3) immediately upstream of the hydraulic jump. The dashed curves (bottom) are calculated values of the flow velocity in the diverging section of the channel (region 4) immediately downstream from the hydraulic jump. The horizontal dashed line at 9 m/s indicates the velocity at which larger boulders at Crystal Rapids can probably be moved by the current. (d) Calculated Calculated values of velocity in region 2, the constriction. The flow is subcritical where the curves are dashed. (e) Calculated decelerations through the hydraulic jump that separates regions 3 and 4.

too severe, the ambient head of the flow, H_r, may not be sufficient to allow the discharge to be accommodated through the constriction. In such cases, critical conditions occur in the constriction, and a *backwater* is required upstream. The deepening of the backwater increases the specific head of the flow over that in the unconstricted part of the channel and permits a greater discharge per unit area through the constriction. The calculated backwater head, H_b, compared to the ambient river head, H_r, is shown as a function of discharge, Q, and constriction value in Fig. 10a. Note from this figure that, for a constriction of 0.25, a backwater head is required for all discharges over about 300 m³/s ($\sim$ 10,000 cfs)[17]. This means that supercritical flow will occur in the rapid at discharges above this value.

Solutions of the shallow-water flow equations (Figs. 10 and 11) show that Crystal Rapids went through the entire spectrum of nozzle behavior shown in the sketches of Fig. 2 as discharges increased during the 1983 flood[18]. At discharges below about 300 m³/s ($\sim$ 10,000 cfs) the flow was essentially subcritical[19] (curve A in Fig. 11). At higher discharges the flow became critical, and then highly supercritical (curves B-E in Fig. 11).

A hydraulic jump is required in the diverging section of the rapid to decelerate the supercritical flow and to drop its energy from the backwater head, H_b, back to the ambient downstream head, H_r. Thus, the model suggests that the large wave that stood in the diverging section of Crystal Rapids at high discharges can be interpreted as a normal hydraulic jump arising from the severe constriction of the channel (Fig. 2). In hindsight, this wave can be recognized as having been present all through the 1966-1983

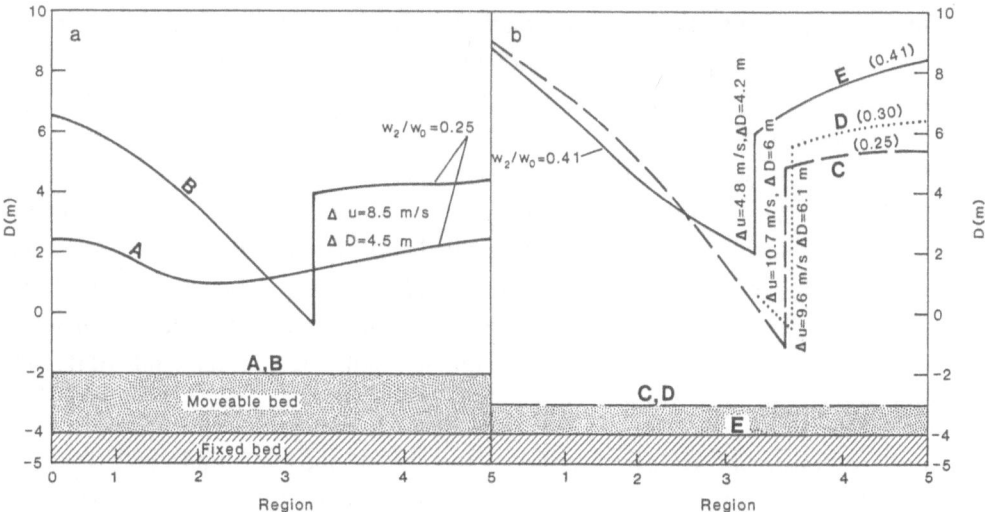

FIG. 11. Schematic longitudinal water profiles at Crystal Rapids for the 1983 discharges up to 2,550 m³/s (90,000 cfs), showing the effect of channel widening and bed erosion on the height of the hydraulic jump. The parts of the rapid defined in the text and in Fig. 9 (regions 0-5) are shown schematically by the labels below the graph. Each curve represents a different discharge and may be related to a different bottom level, depending on the erosion presumed to have taken place. For example, curves A and B are for discharges of 283 m³/s (10,000 cfs) and 850 m³/s (30,000 cfs), respectively, and show the water surface relative to the top of the movable bed labeled A,B. Curves C, D, and E are for discharges of 1,400 m³/s (50,000 cfs), 1,700 m³/s (60,000 cfs), and 2,550 m³/s (90,000 cfs), respectively. Each curve shows the water surface relative to the bed labeled with the same letter; the base level of the bed changes because of erosion, as discussed in the text. Each curve is labeled with the constriction assumed to apply during the 1983 discharges. For the conditions under which supercritical flow occurs (curves B-E), the height of the hydraulic jump and the velocity change across it are given beside the vertical line representing the jump.

phase of Crystal Rapids. However, because the large rock in this vicinity acted like a small-scale natural weir, the energy change of the flow around the rock contributed substantially to the energy of the wave at low discharges, so that the role of the constriction was not recognized. Only when the wave strengthened with increasing discharge, rather than washing out, was the role of the constriction recognized.

As the discharge through Glen Canyon Dam rose from 850 m³/s (30,000 cfs) to about 1,700 m³/s (60,000 cfs) in June, 1983, the height of the wave increased, as would be expected for a hydraulic jump in a channel of fixed geometry (Fig. 10b; compare the left four data bars with the heavy curve). At higher discharges, however, the height of the wave *decreased*, rather than increasing as predicted by the calculations (Fig. 10b; compare the right data bar with the heavy curve).

This puzzling observation can be explained if the magnitudes of the flow velocities in the constriction and upstream of the hydraulic jump are examined (Figs. 10c and 10d; see also the profiles in Fig. 11). In supercritical flow, water accelerates in the converging section of the nozzle, reaching critical velocity, u_2, in the throat. The water continues to accelerate out the diverging side of the constriction, reaching a maximum velocity, u_3, immediately upstream from the hydraulic jump. A sudden deceleration to velocity u_4 occurs across the hydraulic jump as the flow deepens. For example, at a discharge of 1,400 m^3/s (50,000 cfs), with a constriction of 0.25, the calculated velocity in the constriction, u_2, is 9 m/s and the velocity increases to $u_3 = 14$ m/s just upstream from the hydraulic jump (Figs. 10c and 10d). The velocity decreases to $u_4 \sim 3$ m/s just downstream from the jump.

Consideration of the Hjülstrom criterion for particle movement and of unit stream power shows that water moving at 9 m/s can move boulders that are 1-2 m in diameter, the characteristic size of the large boulders of the Crystal debris fan. Therefore, when velocities reached this magnitude (at discharges in the range of 1,400 to 2,000 m^3/s (50,000 to 70,000 cfs), as shown in Fig. 10), large scale erosion began; that is, the river was able to begin contouring its own nozzle. Material was eroded from the sides of the channel and from the river bottom. Vertical erosion scoured the channel in an upstream direction (headward erosion); lateral erosion increased the width of the throat. An observer standing on the shore could not see this erosion taking place, but could hear loud, bass booms as boulders moved in the current.

Channel widening at the throat can account for the observed decrease in height of the hydraulic jump. Comparison of the observed wave height with that predicted for a normal hydraulic jump as the discharge changed from 1,700 m^3/s (60,000 cfs) to 2,600 m^3/s (92,000 cfs) suggests that the channel widened from a shape parameter of 0.25 to about 0.40-0.42, a widening of 12-13 m (Fig. 10b). The location of this erosion is shown in the photograph of Fig. 12.

In summary, a fascinating, and often tense, feedback process involving meteorology, river hydraulics, and engineers began in June 1983 and continued into early July as the discharge increased; this process can be followed on the curves of Fig. 10. As snow melted in the Rocky Mountains, engineers raised the discharge through the dam higher than the 850 m^3/s (30,000 cfs) released during the previous two decades. At the tightly constricted spot in Crystal Rapids, a large hydraulic jump formed because the flow became highly supercritical. As the discharges approached 1,700 m^3/s (60,000 cfs), the river began eroding its channel through the Crystal Creek debris fan. In response to the widening, the flow velocities decreased. If the discharge had been held constant, the channel and hydraulic features of the flow would have stabilized when the channel became wide enough to reduce flow velocities in all sections below about 9 m/s. However, more snow melted in the headwaters of the Colorado River, and engineers were forced to increase discharges through the dam toward 2,600 m^3/s (92,000 cfs). In

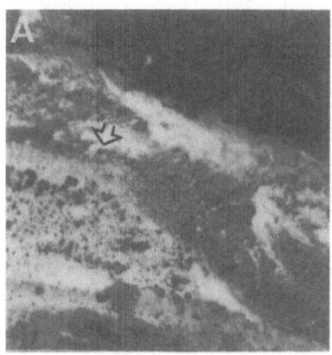

FIG. 12. (a) and (b) Comparison of the shore line at Crystal Rapids (a) before and (b) after the 1983 high discharges. The arrow indicates a large rock visible for reference in both photos. The discharge in (a) is about 283 m³/s (10,000 cfs); in (b) it is about 170 m³/s (6,000 cfs). Note the widening of the channel immediately downstream (right) of the indicated rock. Even though the discharge is lower in (b) than in (a), boats can be seen in an "alcove" in (b) which they could not have reached under the conditions present in (a). The shore where this erosion has taken place is the channel boundary along regions 2 and 3 where the flow velocities were the highest during 1983. In the field, a cut bank nearly 2 m in height can be traced from the left side of (b) to the boats.

response to the increased discharge, flow velocities again increased, and erosion of the channel continued; by the time of peak discharge, enough lateral erosion of the channel had occurred that the height of the hydraulic jump had decreased. It is not clear at this time whether the high flows were sustained long enough for the channel to take on a shape in equilibrium with the high discharge.

C. Implications for geomorphic evolution in the Grand Canyon

Even after the channel was widened by the high discharges of 1983, the constriction of 0.40-0.42 at Crystal Rapids is still significantly below the value of 0.5 characteristic of the mature debris fans along the Colorado River (Fig. 6), and the rapid is significantly different in hydraulic character from rapids at locales where the constriction is 0.5. This observation suggests that most debris fans in the Grand Canyon have been subjected to floods larger than the 1983 flood. With proper recognition of the simplicity of the model and the paucity of data, extrapolation of calculations at Crystal Rapids can be used to estimate the magnitude of flood that might have been required to enlarge the constrictions to the value of 0.5 observed for most debris fans (Fig. 6). A flood of 11,000 m³/s (400,000 cfs) is estimated[16]. This is not an unreasonable estimate, because it is known that a flood of 8,500 m³/s (300,000 cfs) occurred in 1884.

The calculations also indicate that when constrictions of the Grand Canyon debris fans reach the value of 0.45, the flow will be essentially subcritical[19] at all discharges. Although some standing waves and local regions of supercritical flow exist in most of the rapids of the Grand Canyon because of smaller-scale changes in bed elevation (including large rocks) than are considered in this simple model, the wave at Crystal Rapids was unique: other rapids that are less tightly constricted do not have strong normal waves (river rafters might disagree, because many of the existing waves are strong enough to flip rafts, but, at the scale of convergent-divergent constrictions considered here, these are local features.)

Over geologic time, the flow in Crystal Rapids can change from subcritical to supercritical (or vice versa) in two ways: (1) *as discharge changes* from season to season in a channel of fixed constriction, flow may pass from subcritical to supercritical, or vice versa; or (2) *as channel constriction changes* with time because of erosion during large floods or because of tributary debris flows. This evolution is summarized in Fig. 13, a version of Fig. 2 appropriate to Crystal Rapids that shows explicitly the response of channel shape and flow structures to changes in discharge. The sequence shown represents but one cycle in recurring episodes in which debris fans are enlarged by floods in the tributaries and then modified by floods in the main channel.

The beginning of the sequence is arbitrarily chosen as a time when the main channel is relatively unconstricted (Fig. 13a). The river is suddenly disrupted and ponded by catastrophic debris-fan emplacement (Fig. 13b), forming a "lake" behind the debris dam. The surface over which water pours across the freshly emplaced debris fan is called a "waterfall" in this model. As the ponded water overtops the debris dam, it erodes a channel, generally in the distal end of the debris fan (Fig. 13c). This is the beginning of evolution of a "rapid" from a "waterfall." Observations of naturally emplaced earth dams suggest that the breaching of the Crystal debris dam probably happened within hours or days of its emplacement.

Unless the debris dam is massively breached by the first breakthrough of ponded water (that is, unless enough material is removed so that the shape parameter is initially greater than 0.5), the constriction of the river is initially severe. Floods of differing sizes and frequency erode the channel to progressively greater widths (Figs. 13c, 13d, and 13e). Small floods enlarge the channel slightly, but constricted, supercritical flow is still present (e.g., as in Crystal Rapids from 1966 to 1983). Moderate floods enlarge the channel further, and the geometry observed at any instant reflects the largest flood in the history of the fan. Rocks from the debris fan are transported as far as 1 km downstream by the high-velocity water in the convergent, constricted, and divergent regions to form the "rock gardens" that lie below many such debris fans (Fig. 5a shows the Crystal rock garden).

Mature debris fans that once blocked the Colorado River and now have channels cut through them have progressed with time from the bottom to the top of Fig. 2 (or,

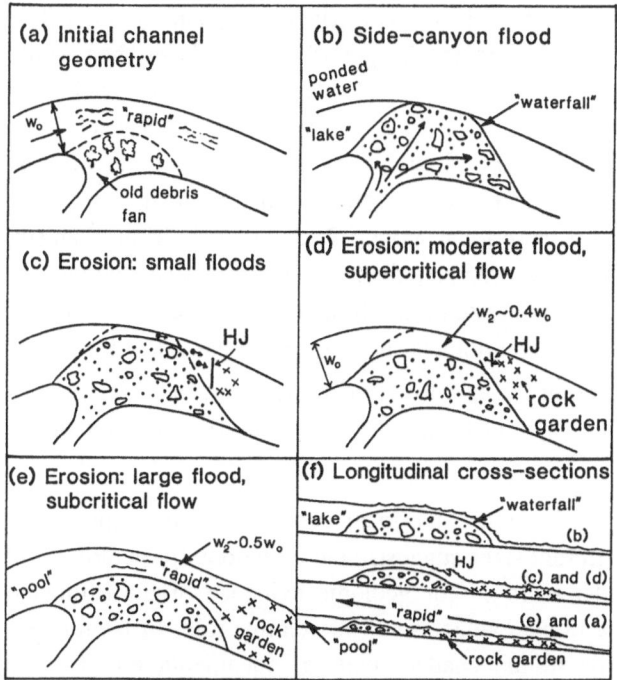

FIG. 13. Schematic illustration of emplacement and modification of debris fans in the Grand Canyon, modeled after the processes observed at Crystal Rapids during 1966-1983. HJ indicates a hydraulic jump. See the text for a discussion of (a)-(e). (f) Schematic longitudinal cross sections through the main channel in (a)-(e).

cyclically, from Fig. 13a to 13e), because natural floods have been large enough to create subcritical channels from the initially supercritical constrictions. Controlled discharges through Glen Canyon Dam will not permit Crystal Rapids to evolve to the configuration of the older debris fans. It is awesome to realize, however, that the controlled releases from the dam (2,600 m^3/s (92,000 cfs)) are sufficient to significantly alter the channel of the river should it become blocked or tightly constricted by tributary floods in the future.

What perspective does this interpretation of the history of the hydraulics and geomorphology of the Colorado River give us on the role of fluid mechanics in geology? First, the interpretation of a major wave as a hydraulic jump arising not from a rock but from large-scale channel geometry provides a different perspective for monitoring and predicting events at newly formed rapids -- a perspective valuable for National Park Service officials concerned with navigational safety. Any rapid newly formed by a debris flow may exhibit hydraulic features different from those seen in channels through older debris fans, because of the severe constriction that can be present. These rapids should be monitored closely if unusually high discharges are put through Glen Canyon Dam.

Second, the interpretation of critical flow in the constriction and supercritical flow downstream from the constriction predicts erosion in quite different places than would be found in a subcritical nozzle, and the interpretation of this shape allows modeling of sizes of ancient floods. Interpretation of the flow in this way also allows a mechanism for transporting large boulders a significant distance downstream from the original debris deposit into rock gardens, because of the high velocities that can occur downstream of the constriction in supercritical flow. Without such a mechanism, geologists are faced with the dilemma that tributary debris flows from side canyons are building weirs across the canyon, and downcutting through these weirs is difficult because only local abrasion or chemical solution can be invoked to get rid of the rocks.

The relative roles of local abrasion, solution, and downstream transport of large particles must be understood before quantitative models of infilling versus downcutting of the Grand Canyon can be formulated.

Finally, the framework of supercritical to subcritical evolution of the rapids with time and through flood events of different sizes suggests new directions for geologic and hydraulic observations at the rapids: searches for geologic evidence of the estimated 8,500 m^3/s (400,000 cfs) prehistoric flood; evaluation of the relative roles of lateral constriction versus vertical topography on the channel bottom; documentation of wave behavior as discharge changes; laboratory experiments on the three-dimensional shapes of hydraulic jumps in flumes of converging-diverging geometry; development of criteria for transport of large boulders (>1 m) by fluids (a research topic that will be discussed again in the concluding section of this paper on the Mount St. Helens lateral blast); and evaluation of the relative frequencies of tributary versus main stem floods in determining the rate of downcutting by the Colorado River within the Grand Canyon.

IV. Old Faithful geyser: a two-phase nozzle

A. Geologic setting

Reports of geysers and hot springs in the land that is Yellowstone National Park began in the early part of the 19th century. Old Faithful Geyser has since become a familiar symbol of the western lands of the United States and their national parks (Fig. 14). More recently, Old Faithful became an important focus of scientific studies when striking resemblances between a peculiar type of volcanic seismicity known as "harmonic tremor" and the seismicity of the geyser were noted and explained[20] (Fig. 15). Harmonic tremor is a relatively monochromatic seismic motion that often precedes or follows volcanic eruptions; as a precursor, it has been invaluable in forecasting eruptions, even though no theory has adequately explained its origin. Harmonic tremor is also an important component of the seismic noise characteristic of geothermal fields, and it is thus potentially an important prospecting tool for geothermal energy sources -- sources that may provide only a few percent of the total energy requirements of the United States, but that could provide a substantial and critical portion of the energy

FIG. 14. Old Faithful geyser, Yellowstone National Park, in eruption. The column is about 30 m high. Note the discrete elements of fluid in the eruption column. These are the "surges" referred to in the text. Photo by G. Mendoza.

requirements of many countries surrounding the Pacific Basin. In this section, I summarize my observations and theory for the origin of harmonic tremor at Old Faithful, the importance of fluid properties in interpretation of the data, and my ideas about the complex process that occurs when the geyser erupts.

For perspective, fluid dynamicists might imagine Old Faithful as a vertical, open-ended, two-phase shock tube with variable cross section. In addition to the compressibility effects that normally dominate shock-tube dynamics, gravitational (hydrostatic) effects strongly influence the fluid properties in the geyser, because each meter of liquid water (providing approximately 0.1 bar pressure) changes the boiling temperature of the fluid by about 2 °C. Thus, in contrast to a shock tube in which initial conditions are usually isothermal and isobaric, the initial pressure and temperature conditions in Old Faithful are not uniform. The recharge cycle of the geyser is analogous to the process of filling a shock-tube driver section with a volatile liquid -- such as liquid Freon -- that can boil upon decompression, except that there is no physical diaphragm to contain the fluid; the natural diaphragm is the highest water in the conduit that maintains a temperature of 93 °C (the boiling temperature at the 2200-m elevation of Old Faithful) and sufficient pressure to keep deeper water from massive boiling and eruption. And

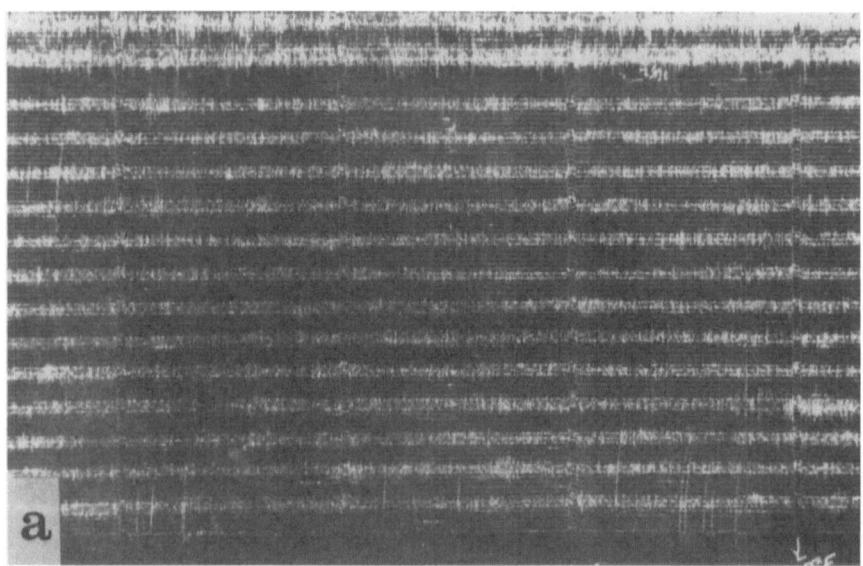

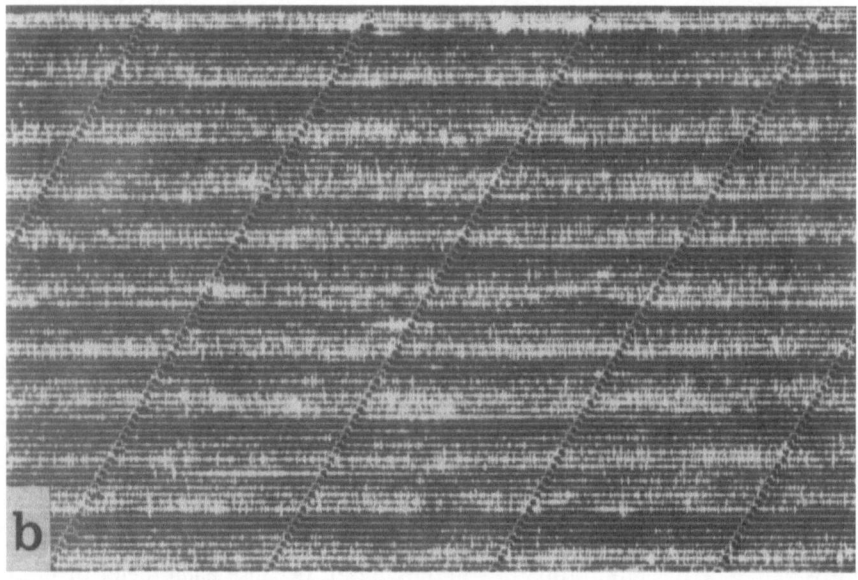

FIG. 15. (a) A seismic record from Karkar volcano, Papua New Guinea (courtesy of C. McKee and R.W. Johnson[21]). The white bands are strong seismic activity at a frequency of 2-4 Hz that recurs at intervals of about 70 minutes. (b) A seismic record of about one day of eruptions at Old Faithful, showing 11 complete eruption cycles (a higher-resolution record of one complete eruption cycle is shown in Fig. 20). The seismometer that obtained this record was a few tens of meters from the vent. Note the similarity to the Karkar record. The time marks in both records indicate 1-minute intervals. Each active period of the seismicity of Old Faithful ends with an eruption and a characteristic eruption coda; a good example of an eruption coda is on line 11,

beginning between the first and second time marks and extending to the fourth. A similar coda at the end of the active times on the Karkar record suggested to McKee and Johnson that underground eruptions were occurring at Karkar.

whereas in laboratory shock-tube experiments we worry about the "cleanliness" of eruption initiation over time scales of micro- to milliseconds, at an eruption of Old Faithful we may be lucky to forecast the initiation time to within 10 minutes!

Study of this complex shock tube must be done under very restricted conditions. Whereas a major problem in the study of the Colorado River discussed in the preceding section is inaccessibility, an equally major problem in studying Old Faithful geyser is its accessibility and public visibility. Observations close to the vent must be made on the few days of the year when work will not detract from tourists' enjoyment of the geyser (namely, when Yellowstone Park is closed for snow-plowing of the roads in late winter), and experiments or observations must be designed to avoid even the slightest damage to the geyser (for example, no hole can be dug to allow positioning and anchoring of a seismometer). Thus, the challenge in studying Old Faithful is to learn as much as possible about the inner workings of a complex nozzle from very limited observations. The basic data set consists of float and thermocouple measurements[22] made in 1949, seismic and movie data[20] taken from 1976 to 1984, and unpublished pressure and temperature data[23] that J. Westphal and I obtained in 1983 and 1984.

B. The recharge and eruption cycle

Old Faithful erupts into a tall, continuous vertical jet of water and steam (Fig. 14). The maximum height of the geyser ranges between 30 and 50 m, depending mostly on wind velocity and, to a certain extent, humidity (which affects the visibility of vapor in the eruption). Eruptions last for 1.5 to 5.5 minutes and occur every 40 to 100 minutes (Fig. 16). Measurements of total discharge suggest that about 0.114 m^3/s (114 kg/s; 1800 gal/min) are erupted during the initial and steady-flow stages of the eruption (E. Robertson, U.S. Geological Survey, private communication, 1977).

The conduit of the geyser, or the nozzle, is a fissure that is flared at the surface, narrows to a constriction at about 3 m depth, and probably diverges into one or more caverns below this depth (Fig. 17). About 0.5 m below the rim of the geyser, the fissure is 1.52 × 0.58 m, and I will take these as the dimensions of the exit plane of the fluid because the actual geyserite surface is very irregular (including a petrified tree stump visible as the knob on the left side of the cone in Fig. 14). Probe work by J. Westphal, R. Hutchinson and me (unpublished data) suggests that the constriction dimensions are approximately 0.1 × 1.5 m. The depth of the conduit that can be reached by a probe, the *immediate reservoir*, is 22 m, although it is plausible that water from greater depths is ejected during a long eruption. Water fills the conduit only to within 6 m of the surface

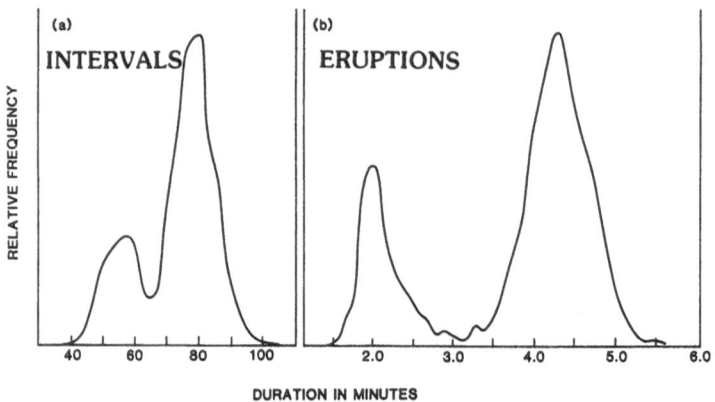

FIG. 16. (a) The relative frequency of intervals between eruptions of Old Faithful. (b) The relative frequency of eruptions of various durations. All data are for the year 1979; the vertical scale is arbitrary. Unpublished data provided by R. Hutchinson, U.S. National Park Service, from analysis of 3,308 eruptions, 1983.

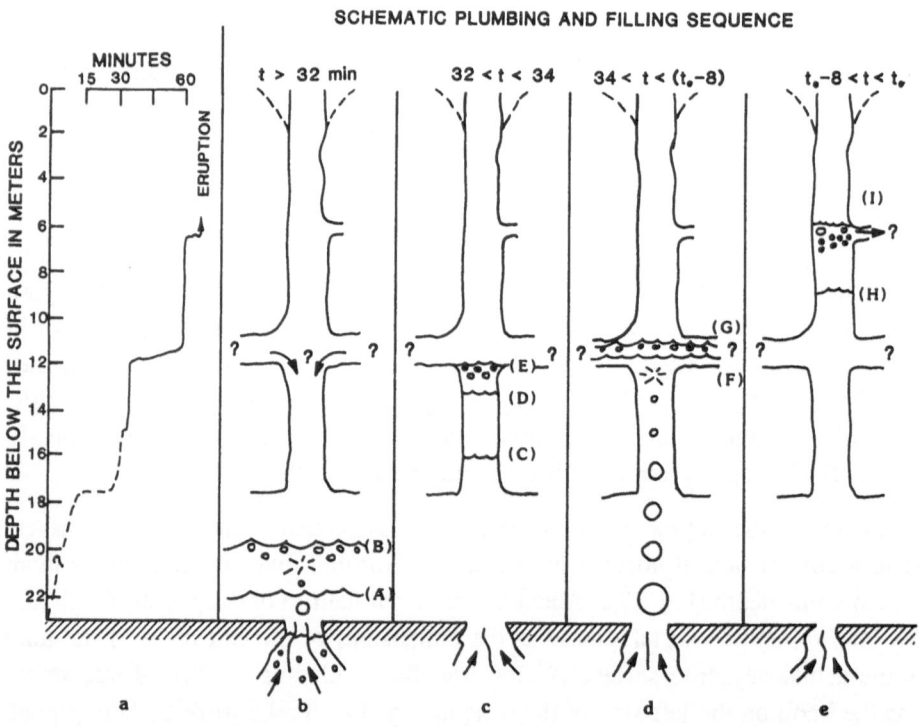

FIG. 17. (a) Rate of filling of Old Faithful, as inferred from float measurements reported in 1948[22]; time = 0 was taken when probe was lowered. Recent work by Kieffer and J. Westphal (unpublished) shows differences in details of the filling rate but does not change the general discussion chosen for this paper. (b)-(d) Schematic diagram of the conduit of Old Faithful, based on the float measurements and an assumed

constant rate of recharge. The time until the next eruption is indicated at the top. The important points to note from these sketches are (1) the gradual recharge of water over the interval prior to eruption (the levels A-I represent inferred depths of the surface water at different times); (2) the addition of heat via hot water and/or steam bubbles at the base of the recharging column; and (3) continuous boiling of the near-surface water. Collapse of steam bubbles within the liquid zone is inferred to be the cause of seismicity.

(Fig. 17). The maximum length of the water column in the immediate reservoir prior to an eruption is therefore about 16 m.

For nearly a century after its discovery, Old Faithful maintained a fairly regular pattern of eruptions, with intervals between eruptions averaging 60 to 65 minutes. During most years, Old Faithful exhibited two types of eruptions: "shorts," which were 2.5-3.5 minutes in duration, and "longs," which were about 5 minutes in duration. During the 1970's, the length of the repose interval, I, following an eruption was quite closely related to the duration of the eruption, D. The empirical formula

$$I = 10 D + 30 \qquad (I, D \text{ in minutes}) \tag{17}$$

proved very useful to the National Park Service for predicting when eruptions would occur.

In the past few years the duration-interval behavior has changed dramatically, although changes in the observable eruption characteristics of the geyser during an eruption (such as height versus time) have not been documented. Intervals averaged over a month commonly exceed 75 minutes and individual intervals have sometimes exceeded 100 minutes. The interval-duration equation no longer applies to the same statistical accuracy; in 1987, short eruptions ceased for a while, and only long eruptions occurred (R. Hutchinson, National Park Service, private commununication, 1987). The mysteries of geyser eruptions have long intrigued scientists (the original theory of the inner workings of geysers was published by Bunsen in 1846), but questions about the inner workings of Old Faithful have taken on a new urgency because of these dramatic changes in behavior.

C. The recharge process: clues to geothermal seismicity

After an eruption ceases, the conduit is empty (or nearly so) and must be recharged with both water and heat. Estimates of total volume erupted and conduit dimensions give an approximate recharge rate of 6 kg/s (liquid water) (see Ref. 20, p. 66). A working model for recharge of fluids and heat to the geyser is based on measurements of depth and temperature versus time[22]. Water rises slowly up the conduit during the recharge interval (Fig. 17). During the rise, temperatures range from 93 °C at the surface to 116°C-118°C at the bottom of the immediate reservoir (Fig. 18, temperature-depth-time curves). The hottest water at the bottom is about 7-9°C below the boiling temperature for the total pressure at the bottom (0.08 Mpa (0.8 bars) atmospheric pressure + 0.14 Mpa (1.4 bars) hydrostatic pressure). Because the deep water is hotter than the shallow water, heat for the recharge cycle is most likely supplied by the addition of hot water or steam at the base of the immediate reservoir.

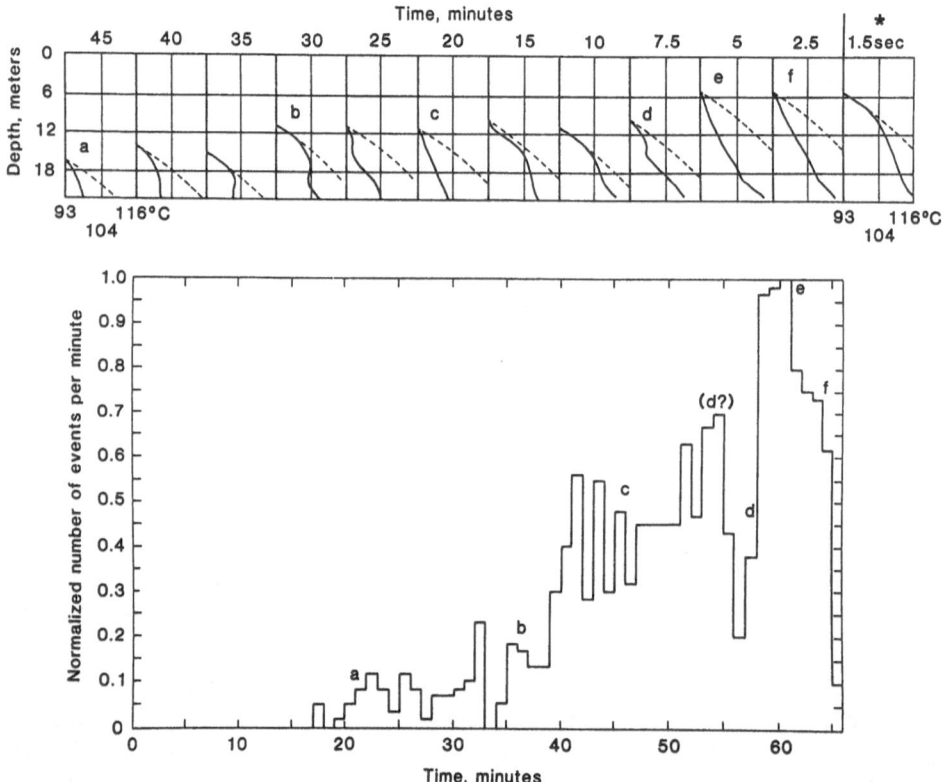

FIG. 18. A histogram of the number of seismic events per minute through a recharge interval at Old Faithful, supplemented by data on the depth of water in the conduit at various times (the graphs above the histogram) and the temperature of the water relative to the reference boiling curve (the top curve in each graph). The temperature-depth data were not taken during the same eruption as the seismic data; for details of the construction of the correlations shown here, see Kieffer[20]. On the bottom axis, time = 0 is taken at the beginning of an eruption that was about 4.5 minutes long; a 66-minute interval followed before the next eruption. The first appreciable seismicity starts at about 21 minutes (45 minutes before the next eruption), and the associated temperature-depth conditions are indicated by the left box labeled "45 min." Successive graphs labeled with decreasing times (40, 35, 30 min, . . . to 1.5 sec) show the gradual filling and heating of the geyser and the correlations with seismic details.

Two processes probably contribute to the mixing of the hot deep water with the cooler surface water: convection and migration of steam bubbles. It is not clear that the two processes can be distinguished using available measurements (note the erratic temperatures at any fixed depth; e.g., at 18 m in Fig. 18). Much of the deeper water is too cool to boil under the total pressure (hydrostatic + atmospheric) at any given time; compare the measured temperatures with the reference boiling curve given in each box in Fig. 18. However, since the temperature of the deeper water exceeds the atmospheric boiling temperature of 93 °C, this water is superheated relative to the boiling point at atmospheric pressure. If such superheated water is convected upward, steam bubbles form as the pressure decreases below the saturation pressure of the water. For example, 3 percent of the deep water at 116 °C will transform to steam when the total pressure decreases to 0.1 Mpa (1.0 bar). As steam bubbles rise, however, they encounter cooler water and may collapse. Although this process cannot be directly observed in the depths of Old Faithful, it is easily observed in a pot on a stove, as well as in other geysers, such as Strokker in Iceland, where the steam bubbles rise into a diverging surface pool and can be observed both to collapse before reaching the surface, and to reach the surface and explode into a beautiful fragmenting shell (Fig. 19). In geysers where the bubbles can be directly observed, they frequently occupy the full diameter of the conduit, which may be of the order of, or more than, 1 m. Collapse of the bubbles, and release of their latent heat, is probably a major process by which heat is transferred upward in the water column[24], and the collapse of these bubbles is believed to cause the individual seismic events observed.

The collapse of a vapor bubble in a liquid of its own composition can occur within milliseconds[25], and pressures in the collapsing cavity can be as large as a few to tens of

FIG. 19. Bubble erupting from the vent of Strökkur geyser, Iceland, through a surface pool of water. The bubble diameter is about 2 m. Note the fine-scale structure on the bubble surface. Photo by H. Kieffer.

megapascals (tens to hundreds of bars)[25,26]. The high pressures generated by the bubble collapse decay quickly with distance, becoming seismic-level disturbances at distances of only a few bubble radii.

The acoustic noise of collapsing steam bubbles can be detected by seismometers placed around Old Faithful (Figs. 15a and 20). The seismic codas of two eruptions, labeled A and B, are shown in Fig. 20. These codas are primarily due to water falling back from the top of the erupting column onto the ground (this fallback was just beginning to occur when Fig. 14 was taken; it is visible on the right side of the photograph). Short, discrete bursts of seismicity occur throughout most of the recharge interval (four such events, indicated in Fig. 20, are shown enlarged in Fig. 21). The number of seismic events per minute increases as the geyser fills and the fluid becomes generally hotter (Fig. 18, histogram).

Two characteristics of this seismicity are the relatively high frequency content of the individual events (a few tens of Hz) and the characteristic damping time of a few tenths of a second. The envelope of these signals is remarkably similar in shape to those obtained by Hentschel[27] in laboratory experiments on collapsing bubbles, although the geyser signals are much longer in duration (tenths of seconds compared with hundreds of microseconds), presumably because of the much larger size of the geyser bubbles.

A collapsing bubble could cause the observed seismicity by generating waves within the fluid column, as illustrated in Figs. 22a and 22b. These waves set the fluid in the conduit into resonance; that is, the conduit is an organ pipe filled with liquid water and is set into resonance by the occasional (or frequent) collapse of a large steam bubble. Detailed treatment of the disturbance caused by a single bubble as a hydraulic transient[20] accounts for the distensibility of the conduit walls and allows the damping of the signals to be calculated. To first order, the frequency of the oscillation induced is assumed to be that of a closed organ pipe:

$$f = a/4L , \qquad (18)$$

where L is the length of the fluid column in the conduit, and, in this equation, the sound speed, a, is the effective sound speed of the fluid modified for the distensibility of the conduit walls[28]. Whereas a is 1,440 m/s for pure water, it decreases to 1,385 m/s if the distensibility of the walls of the conduit is accounted for.

The length of the fluid column in the conduit, L, depends on the time in the recharge cycle (note in Figs. 17 and 18 how slowly the recharge process occurs): L is estimated to be 8 m when the seismicity is first detected and 16 m when the conduit is full near the time of an eruption (Figs. 17c-17e and Fig. 18). The corresponding resonant frequencies obtained from Eq. (18) are 43 and 20 Hz, values that approximately span the range of measured frequencies during the recharge interval (Fig. 21). From hydraulic transient theory, the characteristic damping time is 0.12 to 0.46 s, in good agreement with the duration of the observed pulses (Fig. 21).

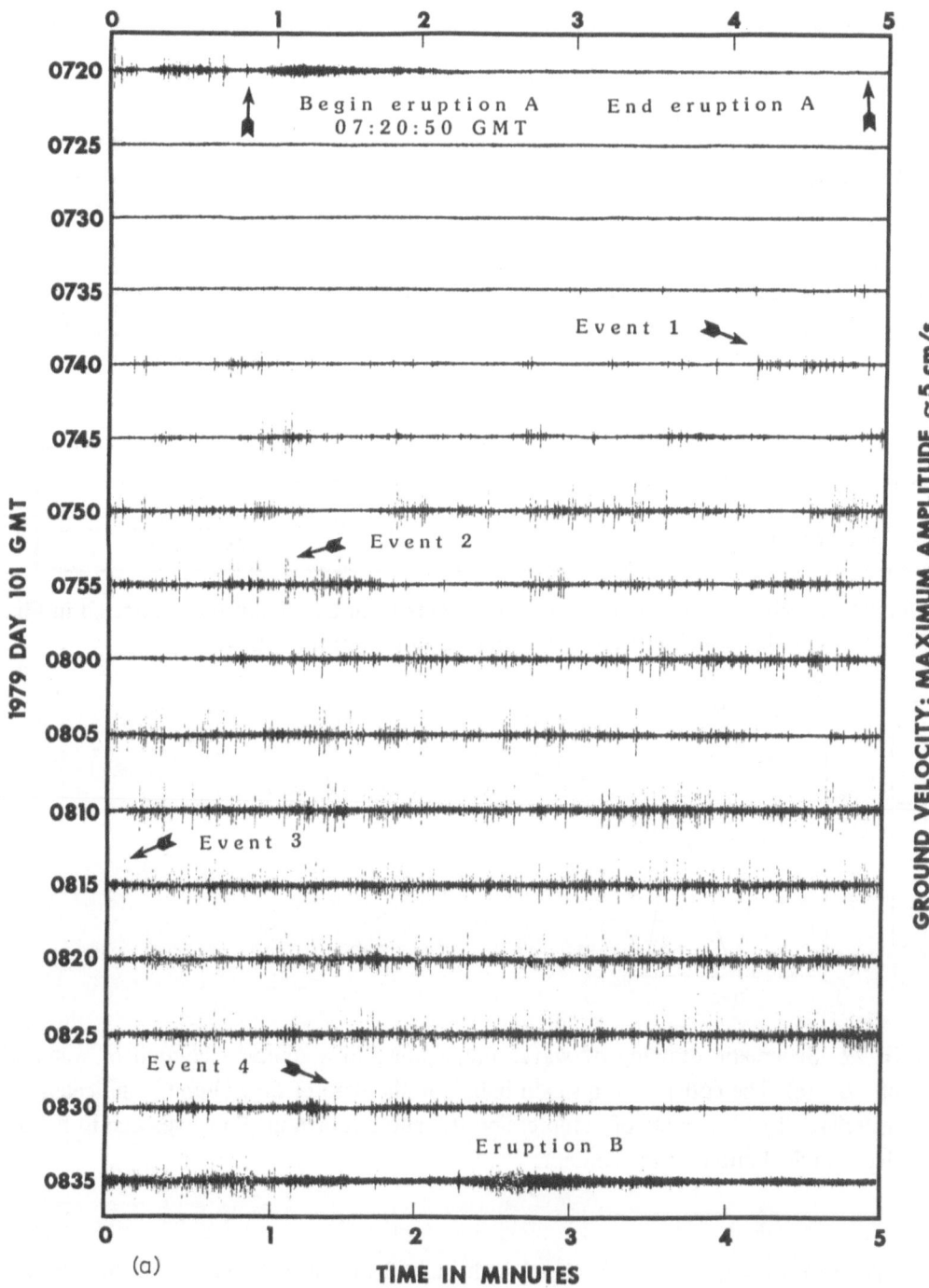

FIG. 20. A seismic record of one eruption cycle of Old Faithful. Details of the four seismic events labeled 1-4 are shown in Fig. 21. To obtain the histogram of Fig. 18, the number of these events exceeding a peak-to-peak amplitude of 4 cm per second was counted.

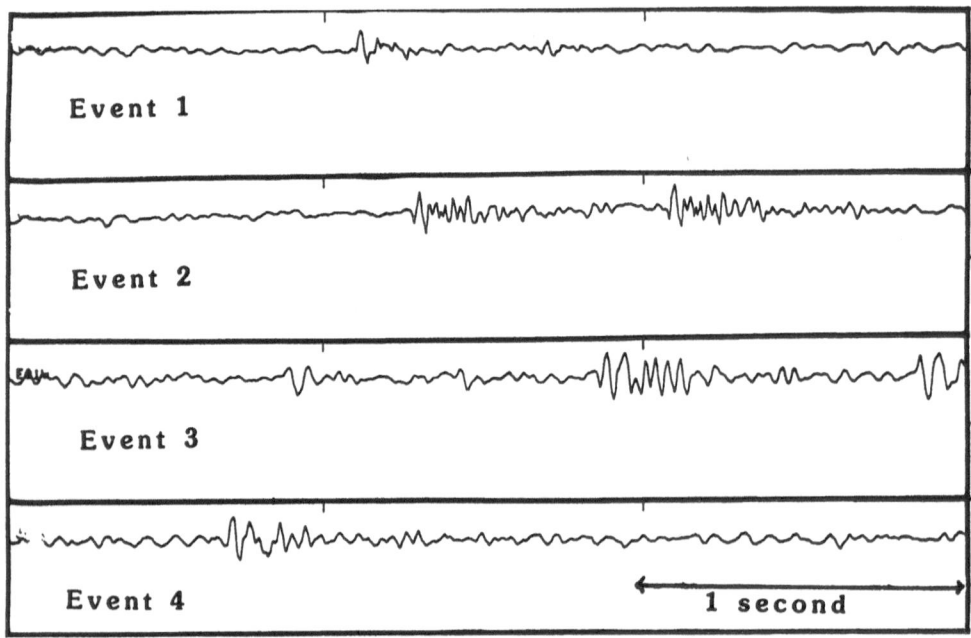

FIG. 21. Details of four impulsive seismic events from Old Faithful as indicated in Fig. 20. Note that the dominant frequencies range from ~ 20 to ~ 40 Hz.

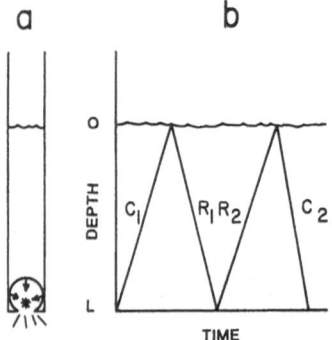

FIG. 22. Schematic drawing of waves propagating in a standing column of water of depth L. (a) The collapse of a steam bubble in the base of the column is indicated by an asterisk. (b) One cycle of compression (C) and rarefaction (R) waves due to bubble collapse in the bottom of the conduit.

The resonant frequency associated with individual seismic events, and the duration of each event during the pre-eruption seismicity at Old Faithful, can be explained, to first order, as arising from hydraulic transients in a slowly recharging column of *liquid* water. If the fluid in the conduit were, for example, a boiling, two-phase mixture, the sound speed would be dramatically lower, and the observed seismicity could not be explained so simply as arising from resonances of the liquid column standing in the immediate reservoir. In the next section, I describe the thermodynamics of the transformation of the recharged liquid in the conduit as an eruption occurs, and show how the associated change in sound speed of the fluid influences the resonant frequencies of the conduit.

D. Eruption dynamics and thermodynamics

An "eruption cycle" of Old Faithful consists of one eruption and one recharge interval. By convention, the start of an eruption cycle is taken as the onset of an eruption, although the first part of an eruption actually begins below ground level and cannot be monitored (see the discussion of preplay below). The visible flow field and its variation with time during an eruption are collectively referred to as the "play" of the geyser. The play consists of four parts which can be distinguished on a graph showing the height of the eruption column versus time (Fig. 23).

The first part of an eruption (the part that begins underground) is *preplay*, the ejection of water intermittently prior to the actual eruption (the last few episodes of preplay before the eruption, from −15 s to 0 s, are shown in Fig. 23). Episodes of preplay last for a few seconds, and water is thrown to a height of a few meters or occasionally a few tens of meters. During preplay, water ejected upward into the atmosphere is cooled by expansion and entrainment. When it falls back into the conduit, it mixes with the heated conduit fluid, and the resulting cooling of near-surface water can delay the onset of an eruption until the fluid has been reheated to initiation conditions.

During *initiation and unsteady flow* the eruption column develops and rises in a series of *bursts* (typically 1 to 8 in number) to a maximum height of about 30 m to 50 m. This part of the eruption is typically 20 to 40 s long. Two minor bursts and four major ones occurred during the first 20 s of the eruption documented in Fig. 23.

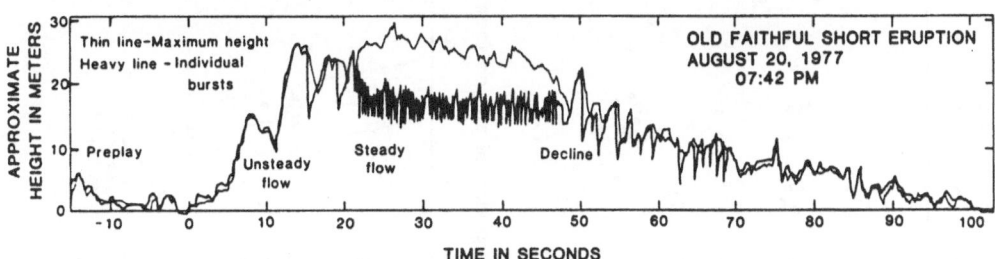

FIG. 23. A record of height versus time for a short eruption of Old Faithful, showing the four stages of eruption discussed in the text. The light line is the height of the water-steam column; the heavier line traces individual pulses of water, visible in Fig. 14, in order to obtain the frequency of surging (approximately 1-2 Hz).

Steady flow is an interval of about 30 s during which the column stays near maximum height. During this time, *surges* are observed in the eruption column at a frequency of about 2 Hz in short eruptions and 1-1.5 Hz in long eruptions (Fig. 24b). The surges are shown photographically in Fig. 14, and graphically from 20 to 45 s of the eruption documented in Fig. 23.

The *declining part of the eruption* begins rather dramatically, after roughly 30 s of steady flow, with a drop in column height. The decline can last up to 3 min, and during this time the height of the column drops to about 10 m and play continues at low levels.

Differences between long and short eruptions are threefold: the frequency of surging during the steady-flow stage; the duration of the decline stage; and the seismic pattern following the eruption. Analysis of height-time data (such as shown in Fig. 23) for many eruptions reveals that there is no correlation of maximum height obtained, duration of maximum height, or number of bursts in the initiation stage with eruption duration (Fig. 24a). The only measurable differences in eruption play between long and short eruptions are the frequency of the surges (Fig. 24b) and the duration of the decline phase, which is simply truncated at about 2-3 min for a short eruption. Seismically, a period of about 20 min of quiet follows a long eruption (Fig. 20), whereas seismicity begins immediately after a short eruption. After short eruptions, we are able to hear water splashing in the bottom of the immediate reservoir, which suggests that it is not completely emptied during short eruptions and that the seismicity arises within the immediate reservoir.

Water near the surface of the recharging column boils nearly continuously, and vigorous boiling can accelerate fluid several meters upward. Early in the recharge cycle, the fluid is too deep in the conduit for vigorously boiled water to be ejected over

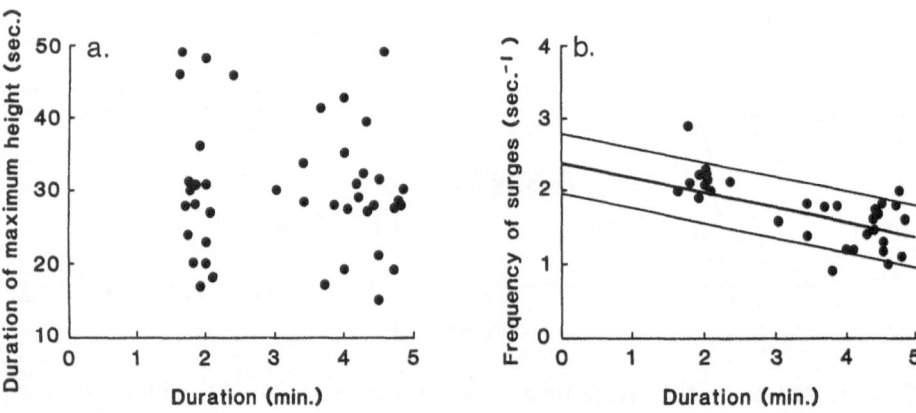

FIG. 24. (a) Graph of duration of steady-flow stage (maximum height) versus duration of eruption of Old Faithful geyser, showing no correlation. (b) Graph of frequency of surges versus duration of eruption, showing inverse correlation.

the rim. Thus, even though vigorous boiling and splashing occur within the conduit, the pressure distribution at depth remains unchanged because fluid is not removed from the conduit and an eruption cannot begin. An eruption therefore begins when two criteria are met: (1) the water must have risen high enough in the conduit so that vigorous boiling can discharge some water over the rim, thereby reducing the hydrostatic pressure on the fluid by removal of some mass; and (2) the underlying water must be hot enough that the mass unloading triggers a positive feedback process.

As the recharging fluid rises to within about 6 m from the top of the conduit, boiling can eject fluid out onto the cone of the geyser. This removal of mass reduces the pressure on the underlying fluid. If the underlying water is sufficiently close to the boiling curve, and if enough water is removed to start the cascading process of decompression, the unloading results in a positive feedback process whereby more boiling, and hence an eruption, occurs. If the underlying fluid is not sufficiently close to the boiling curve -- e.g., if relatively cool surface water has been recently overturned by convection -- the unloading may simply result in a burst of preplay and no eruption.

A simplified diagram of Old Faithful modeled as a layered shock tube is shown in Fig. 25, so that the words "enough" and "sufficiently close" in the above paragraph can be semiquantitatively defined. In the example shown, I have arbitrarily divided the fluid in the column into six cells of different, but uniform, temperature (cells B, C, D, E, F, G); these might be thought of as simplified convection cells. These cells mimic the temperature curve measured by Birch and Kennedy[22] 15 s before the onset of an eruption (Fig. 18, last temperature-depth graph). The top zone, A, 1 m in length, has been assumed to be continuously boiling and to be the mass of water that is ejected out of the conduit to start the eruption (Fig. 25a). I chose the length of 1 m for this zone somewhat arbitrarily after watching the onset of about 200 eruptions and estimating the amount of water ejected.

When A is ejected out of the conduit, the pressure on the underlying water is everywhere reduced by the weight of the ejected fluid (assume this to be about 0.1 bar; SI units are abandoned temporarily here for ease in following Fig. 25). Relative to the reference boiling curve, therefore, the initial temperature distribution curve is elevated toward the boiling curve by 0.1 bar. If it is assumed that no large-scale temperature inversion exists (such as J, Fig. 25), some underlying water is now superheated with respect to the boiling temperature at the new pressure; specifically, any water that was originally within 0.1 bar of the reference boiling curve. In the example shown, the water between B and B' will boil (Fig. 25b). When this water is erupted, more underlying water will boil: any water that was originally within 0.2 bar of the reference boiling curve. In Fig. 25, this is only the water between B' and B'', although the water at the top of cell C is now very close to the reference boiling curve (Fig. 25c). When this water has been unloaded, the pressure is everywhere 0.3 bar less than the initial pressure, and water between B'' and B''', as well as between C and C', will boil. Note

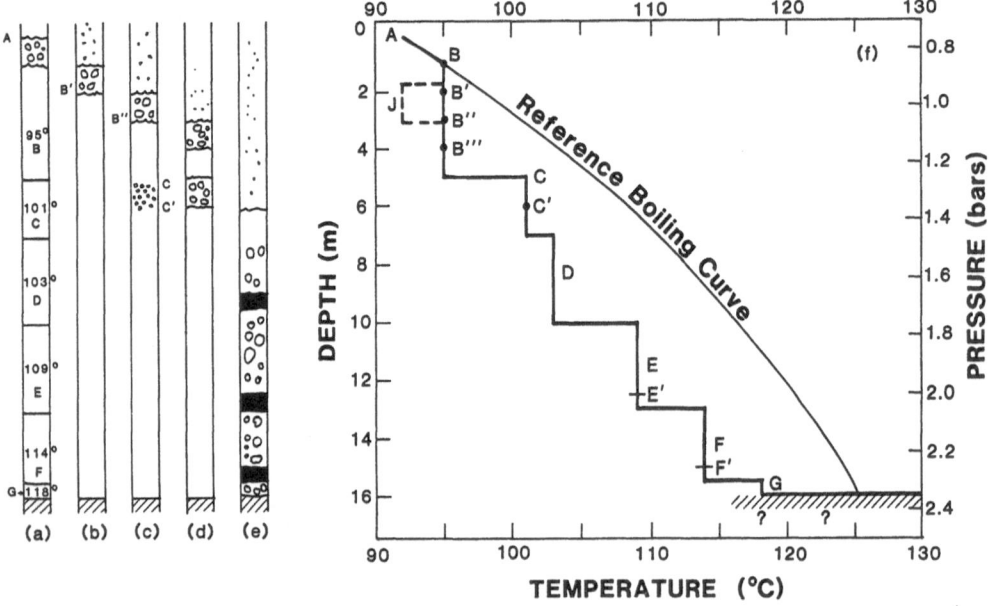

FIG. 25. Old Faithful modeled as a shock tube containing layers of fluid at different initial pressures and temperatures, shown in their position relative to the reference boiling curve in the right figure, (f). In (f) the initial temperature is assumed constant within six convection cells (labeled B, C, D, E, F, and G). The fluid between A and B is assumed to be at pressure-temperature conditions on the boiling curve in order to trigger the eruption. (a)-(e) Five stages in the unsteady unloading process of an eruption of Old Faithful; release of each of the five convection cells would presumably correspond to a "burst" observed as the geyser evolves into steady flow (see Fig. 23). The segments shown with large "bubbles" are boiling; the segments shown as a fine "mist" are erupting from the conduit; and the three segments shaded black are liquid H_2O because of the relations of parts of cells D, E, and F to the reference boiling curve during the unloading process. The thermodynamic path of the unloading process is discussed in the text.

that, because of the weight of B″-B‴, the fluid between B‴ and C cannot boil until some of the fluid in B″-B‴ erupts from the conduit. Presumably, as the water between B″ and B‴ erupts, and as C-C′ boils and expands, the liquid water between B‴ and C will be pushed up and will boil as the pressure decreases. Slugs of liquid water trapped between slugs of boiling froth should be expected because of the different possible relations of the fluid temperature to the reference boiling curve (Figs. 25c-25e).

When fluid has been erupted down to level C′, 6 m of water will have been ejected, and the pressure in the column will be everywhere 0.6 bar less than the initial pressure. Except for small amounts of fluid between E′ and F and F′ and G (Fig. 25e), the fluid will be everywhere on the reference boiling curve, and the conduit becomes nearly completely filled with a two-phase mixture. The fluid is probably a boiling liquid at depth, grading upward into the steamy aerosol that emerges at the surface.

Further details of this unloading process are unknown; depending on the constriction, shock and rarefaction waves may play a prominent role in ejecting the different layers of water[20], or choked flow may be more important in controlling the mass flux rate (see below). Resolution of this question may only arise from detailed theoretical or laboratory modeling, because of the extreme difficulty of field measurements of conduit geometry and time history of the fluid flow.

At the onset of an eruption, liquid water is present everywhere in the conduit except in the boiling zone at the surface. Boiling to progressively deeper levels decreases the amount of liquid present and replaces it with a two-phase mixture (Fig. 25). Because the initial pressure-temperature curve at the onset of an eruption lies no farther than 0.05 to 0.07 Mpa (0.5 to 0.7 bar) below the reference boiling curve, the whole column of fluid in the conduit lies on the reference boiling curve after unloading of only 5 to 7 m of the water. Thus, when about half of the vertical length of the water column has been unloaded, the pre-eruption "organ pipe" filled with liquid water has been transformed into a steamy two-phase nozzle. (Note that if the conduit is perceived as a nozzle, the original standing water only occupied 16 m of length, whereas the erupting fluid fills the full 22 m, and extends into a jet 30 to 50 m high outside the conduit.) It is likely that the "bursting" observed during the unsteady initiation of the eruption represents the eruption of progressively hotter parcels of water, and that the transition from unsteady bursting flow to steady surging flow observed in the behavior of the eruption column (Fig. 23) occurs when the whole immediate reservoir is filled with a two-phase fluid.

As the water in the base of the reservoir is decompressed to atmospheric pressure, some of the enthalpy stored in the hot fluid is converted to kinetic energy. The hottest water, at 116 °C, has an enthalpy (relative to the triple point) of 486.72 kj/kg and an entropy of 1.4842 kj/(kg·K). If this fluid decompresses isentropically to 93 °C, at 0.08 Mpa (0.8 bars) pressure, 4 percent of the liquid is converted to vapor. The final enthalpy of the mixture is 482.83 kj/kg, so 3.89 kj/kg are available for kinetic energy. This is sufficient energy to accelerate the fluid isentropically to a velocity of 88 m/s.

Although the velocity at the exit plane of the geyser has not been directly measured, a simple ballistic calculation of velocity based on the height of the eruption column can be used to estimate ejection velocity[20]. This calculation, using $x_m = u_0^2/2g$, where x_m is the maximum height, u_0 is the exit velocity, and g is the acceleration of gravity, gives $u_0 = 31$ m/s for $x_m = 50$ m. If the acceleration is presumed to begin at the base of the conduit, x_m could be about 70 m, and u_0 from this model is 37 m/s. However, because an eruption of Old Faithful produces a jet rather than a "ballistic billiard ball", the simple ballistic equation may not accurately estimate the exit velocity. Experiments on the dynamics of negatively buoyant plumes[29] give the relation

$$x_m = 1.85 \, F^{1/2} D \ , \tag{19}$$

where F is the densimetric Froude number:

$$F = u_0^2 \left[\frac{\rho_0}{(\rho_a - \rho_0) \, g \, D} \right] . \tag{20}$$

In this equation, ρ_0 is the fluid density, ρ_a is the density of the fluid into which the jet is emerging, and D is the diameter of the jet, assumed axially symmetric. (This F is comparable to $(Fr)^2$, with Fr defined as in Eq. 15). The absolute value of the density difference is used in Eq. (19), where the square root of F is needed. For the exit plane at Old Faithful (1.5 m × 0.6 m fissure), take $D = 1.1$ m. Assume that $\rho_0 = 11.23$ kg/m^3 (4 percent vapor) and $\rho_a = 0.7$ kg/m^3. The above equations then give $u_0 = 78$ m/s, a value in surprisingly good agreement with the velocity of 88 m/s predicted simply from the enthalpy change of the fluid.

The low-frequency surging ($\sim$ 1-2 Hz) observed when the geyser is in steady flow (Figs. 14, 23, and 24) could be the resonances of the conduit filled with the two-phase mixture. The sound speed of an H_2O mixture that is 4 percent vapor is 57 m/s at 1 bar pressure and does not vary significantly between 0.8 bar atmospheric pressure and 0.18 Mpa (1.8 bars) vapor pressure for the fluid boiling at 116 °C at the bottom of the vent (the sound speed would decrease dramatically as the vapor fraction approaches zero, as shown in Fig. 3, vertical line (a), but I assume that the vertical range over which this occurs is small). The resonant frequency of a 22-m closed pipe with this sound speed would be 0.64 Hz, corresponding to a period of 1.5 s. This frequency is about a factor of 2 lower than that measured for a long eruption (1-1.5 Hz), but it probably must be considered in satisfactory agreement, given the simplicity of the model, the unknown geometry of the conduit, and the extreme difficulty of measuring each individual eruption surge.

Because we have no reason to believe that the sound speed of the fluid is any different in long versus short eruptions, the change in surge frequency with eruption duration can be interpreted as representing an effective change in the length L of the resonating column. The possibility that multiple chambers containing fluid exist cannot be excluded, and emptying of different numbers of such chambers could account for an effective change of L. Another possibility is that water is vaporized to different levels in the eruptions of differing durations. During a long eruption, for example, all water in the immediate reservoir is converted into a two-phase mixture, and the immediate reservoir is completely emptied (evidence for this consists of lack of any audible splashing within the reservoir for about 20 minutes after such an eruption, and lack of seismicity during this time). For the long eruptions, the bottom of the conduit is probably truly the geyserite bottom reached by probe, and L is the measured 22 m. During a short eruption, only part of the water in the immediate reservoir appears to be discharged (as evidenced by audible splashing in the conduit immediately after such an eruption ceases, and concurrent resumption of seismicity). There is, therefore, probably a level below which water does not vaporize during short eruptions. In these eruptions, the surface of the unvaporized water would be the effective bottom of the reservoir, because of the large difference in acoustic impedance between boiling and liquid water; that is, the length L would not be the conduit length, but a shorter value equal to the length of water column vaporized. This could account for the higher surge frequencies observed during short eruptions (Fig. 24b). The mechanism whereby the water, which was at 116 °C at the beginning of the eruption, is prevented from vaporizing under pressure reduction remains a mystery. The most common speculation is that cold water can occasionally enter the reservoir during an eruption, but there is no proof of this.

E. Speculations and summary

Is the flow from Old Faithful choked? Is there supersonic flow anywhere in the eruption jet? There are too few data to permit firm conclusions. A few speculations can be offered, although calculation of choking conditions in two-phase flow is a notoriously difficult problem, even in well-designed pipes[30].

The highest vapor pressures in the reservoir will be generated as the hottest water boils -- 0.175 Mpa (1.75 bars) as the 116 °C water boils. Equilibrium expansion to atmospheric pressure of 0.08 Mpa (0.8 bars) produces a fluid that is about 4 percent vapor. Given a maximum reservoir pressure of 0.175 Mpa (1.75 bars), the choke pressure can be calculated from theoretical considerations to be about 0.13 Mpa (1.3 bars)[31]. Experimental evidence[32] suggests that this calculated choke pressure is too high and that the choke pressure could be as low as about 0.55 of the reservoir pressure, about 0.09 Mpa (0.9 bar). The near similarity of atmospheric and estimated choke pressure suggests that the flow could be choked when this hottest water is flowing, but that supersonic flow in the diverging part of the conduit will be weak. The discussion above of the decompression process suggests that choked flow would be most likely during the steady-flow stage, beginning 20 to 30 seconds into the eruption.

Assume that choking occurs at conditions very close to those seen at the exit plane; that is, at 0.1 Mpa (1 bar) pressure when the fluid is about 3 percent vapor. The mass flux is given by

$$\dot{m} = \rho A \, a^*,$$
(21)

where a^* is the sound speed, 45 m/s for equilibrium. The density of the fluid is 19.68 kg/m^3 (3 percent vapor), and the choke area is about 0.15 m^2. The calculated mass flux is 132 kg/s (2100 gal/min). This is satisfactorily close to the measured value (114 kg/s; 1800 gal/min), given the extreme difficulty of measuring the discharge accurately and the relatively large uncertainty of the choke area, A.

If, as calculated above, the exit-plane velocity is ~ 80 m/s, and the equilibrium sound speed at the exit plane is 57 m/s, the implied Mach number at the exit plane is ~ 1.5, barely supersonic within the uncertainties of the modeling. It is therefore not surprising that shock features such as Prandtl-Meyer expansion, visible shock waves, or noise originating from shocks within the plume are not observed at Old Faithful. In contrast, Beehive Geyser (Fig. 26a), which sits just a few hundred yards from Old Faithful and erupts too erratically to be monitored, sounds like a jet engine and, with a little imagination, can be envisioned to contain internal shock waves. These shocks are similar to those observed at weakly supersonic geothermal wellheads (Fig. 26b).

In summary, Old Faithful is a complex two-phase nozzle, possibly sonic or weakly supersonic, and certainly large enough in scale for both gravity and compressibility to

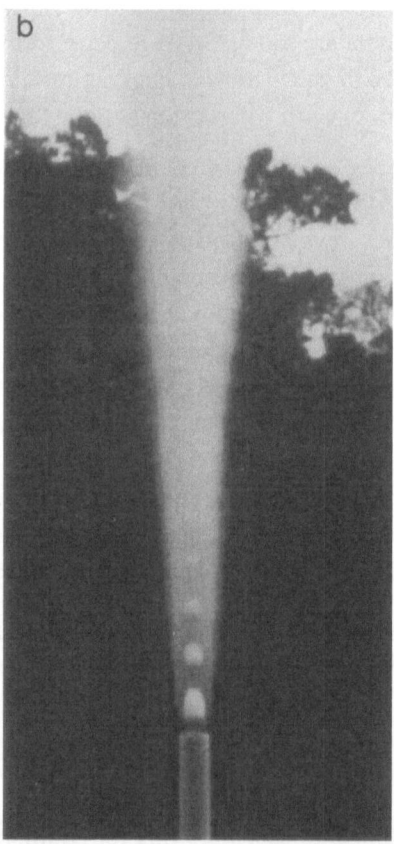

FIG. 26. (a) Beehive geyser in eruption; the cone is about 1 m high. The arrow points toward three white diamonds in the center of the flow, interpreted as shock-wave wave structures within a supersonic flow. Photo by Jeremy Schmidt[33]. (b) Photo of geothermal well MG-5, Tongonan, Philippines. Photo by Charles Darby, KRTA Ltd., Auckland, New Zealand.

be important. Although available data still do not permit a detailed model for the eruption dynamics, they have served to point out new directions for experiments and observations, some of which are now in progress. One of the most important directions of research focused on by these discoveries relates to the similarities in seismicity between geysers and volcanoes that exhibit harmonic tremor (Fig. 15). Harmonic tremor has for decades been attributed to magma motion in volcanoes, but the quantitative nature of the mechanism causing it has been elusive. The geyser study suggests that bubbles in ground water contained in fissures or pockets surrounding hot magma could be the source of tremor[20], and quantitative studies of this mechanism are now in progress[34]. Because we know neither the dimensions of the conduits containing the fluid nor the nature of the fluid that is causing the seismicity at volcanoes, this is a very difficult problem. The study of Old Faithful, where at least constraints can be put on the fluid and on the conduit dimensions, has been important in developing the volcanic ideas. Fluids proposed for the source of volcanic tremor (undersaturated magma, gassy magma, water) can have sound speeds that differ by nearly three orders of magnitude, and the above discussion suggests that there will be an ambiguity in decoupling the effects of conduit dimensions from fluid properties in any analysis of volcanic harmonic tremor. This is an active area of research in volcanology because of the regular occurrence of volcanic tremor at some volcanoes near heavily populated areas where forecasting has enormous implications for life and economy; e.g., at Ruiz Volcano in Columbia. Fluid dynamicists can potentially contribute important ideas and measurements to this problem: laboratory studies of the dynamics of large collapsing bubbles and of two-phase flow in long pipes of variable area are needed. In particular, theoretical and experimental work on the dynamics of compressible flow with gravitational effects will be required to deal with problems involving these fluids at geologic scales.

V. Mount St. Helens: a supersonic jet

A. Geologic setting

Mount St. Helens became famous as an "active volcano" on May 18, 1980. However, that eruption was heralded by nearly six weeks of precursor activity during which eruptions were strikingly similar in scale, frequency, and fluid dynamics to eruptions of major geysers like Old Faithful. On March 27, 1980, after a repose of one century and ominous seismicity for a week, an unobserved eruption created a small crater in the summit of the mountain. For a few weeks, eruptions of steam and ash emanated from the summit (Fig. 27). Studies of deformation of the mountain, and later events, strongly suggest that magma was being intruded into the edifice from depth at this time (Fig. 28a). Water near the magma was heated and convected upward, emerging in eruptions that were geyser-like in scale (hundreds of meters to a few kilometers high), duration (minutes to tens of minutes), and frequency (every few hours). These early eruptions were driven by heated ground water -- no magma was involved.

FIG. 27. A geyser-like eruption of Mount St. Helens, April 1, 1980. An ash-laden density flow rolls down the southwestern slopes (to the right in the photo) from the summit crater at about 2,930 m (9,670 ft) elevation, while steam separates and rises to about 4,500 m (15,000 ft). The fluids in the density current and in the dusty steam might be quite well modeled by pseudogas approximations, because they are relatively dilute. Photo by H. and S. Kieffer.

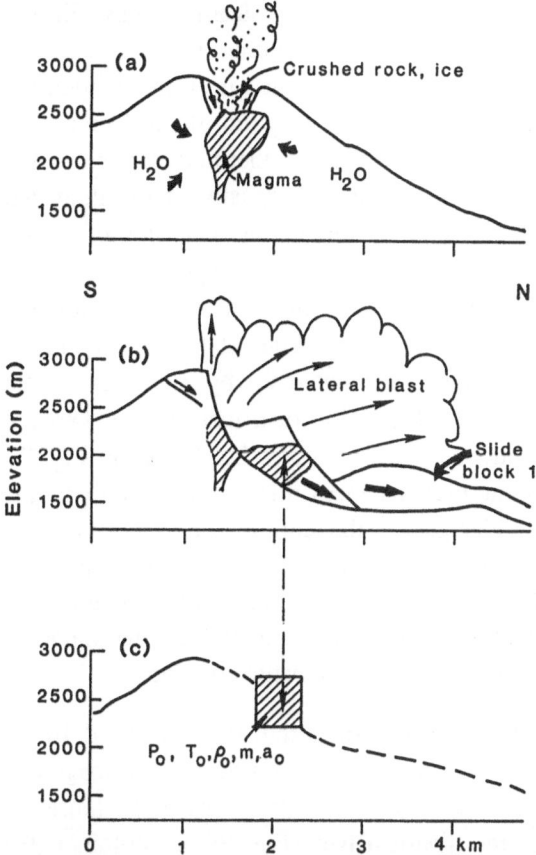

FIG. 28. (a) South-to-north cross section showing schematically the conditions inside Mount St. Helens during March, April, and early May, 1980. Magma has moved high into the mountain. Water heated by the magma has risen through fractures and erupted (a typical eruption is shown in Fig. 27), creating a conduit and summit crater. Crushed rock, ice, ash, and water that intermittently choked the conduit were reworked by successive eruptions. (b) Reconstruction[36] of the initiation of the lateral blast on May 18. The mountain failed along three major faults (indicated by arrows). Magma was present in each, but appeared to emanate mainly from slide block 2 (shown in motion) and slide block 3 (indicated on top of the south part of the mountain). (c) Schematic drawing of initial conditions assumed for the fluid flow model. The complex structure of the landslides has been simplified, and the reservoir is approximated as a single volume whose dimensions are given in the text. In spite of the complicated time history of the landslides, the triggering is assumed to be instantaneous for modeling purposes. The material erupted was partially magmatic and partially hydrothermal. Only the top part of the source shown in (a) and (b) would have been erupted into the lateral blast. The remainder was erupted into a tall vertical eruption column during the several days after the lateral blast. The magma properties are assumed homogeneous at initial pressure, P_0, temperature, T_0, density, ρ_0, and sound speed, a_0 (calculated for a pseudogas with mass loading m).

Although these eruptions were more geyser-like than "volcanic," they differed thermodynamically from eruptions of most geysers because the erupting vapor carried a heavy load of particulate material -- crushed rock and ice gouged from the conduit and crater (this particulate material gives the lower part of the eruption plume in Fig. 27 a dark color). The mass loading by this material affected the thermodynamics in two ways: the entrainment of solid fragments increased the bulk density of the fluid, and heat transfer between the solids and the expanding gas altered the expansion of the gas from that which would be obtained by a two-phase mixture or vapor alone. In the pseudogas approximation discussed in Section II, mass loading is taken into account as an increased molecular weight of the mixture (see the equations at the top of Fig. 4). Heat transfer from hot particles to cooler vapor is accounted for by a decrease in the isentropic exponent of the perfect-gas law. The expansion of a mass-loaded vapor is contrasted with the expansion of a vapor alone, or a decompressing liquid, in Fig. 29. As can be seen from this figure, under some circumstances mass loading can simplify the fluid dynamics by preventing phase changes -- the entropy of a gas phase is increased by heat conducted from solids, and thus formation of a condensed phase is suppressed. However, complications of heat transfer, drag, and interparticle interactions arise. No theoretical models yet handle these effects realistically for the range of particle sizes, particle shapes, and mass loading typical of volcanic eruptions.

The north flank of Mount St. Helens was badly fractured and weakened by the intrusion of magma in March and April, 1980. At 8:32 a.m. on May 18, a magnitude 5.2 earthquake shook the mountain, and several large landslide blocks broke loose and slid downhill toward the North Fork of the Toutle River (Fig. 28b). Within a few seconds, the pressure on magma, hot water, and gases inside the mountain was greatly reduced, and their rapid expansion produced the devastating event that was to become known as the "lateral blast". The evolution of this blast was recorded by several eyewitnesses, by seismic equipment stationed around the mountain, by weather barometers, and by damage to the environment and to man-made equipment around the mountain[35]. Nearly 600 km^2 of forest were devastated, and approximately 60 people were killed. Heavy logging equipment was tossed and overturned (Fig. 30a); trees were totally stripped from the land over a large area. Where they remained, as much as 10 cm of bark and wood was abraded and the interiors were impregnated with shrapnel (Fig. 30b). The pattern of tree blowdown (Fig. 31) provided a remarkable record of local flow directions -- certainly a flow-field pattern to challenge geologists and fluid dynamicists for years to come.

In the region closest to the volcano, trees were either stripped from the land or were felled subradially away from the vent, and the blowdown direction showed little dependence on the terrain (Fig. 32). This zone is called the *direct blast zone*, to emphasize that the blast travelled directly away from the mountain without regard to even major topographic obstacles[37]. Surrounding the direct blast zone is a zone in which topography did influence the blowdown direction, called the *channelized blast*

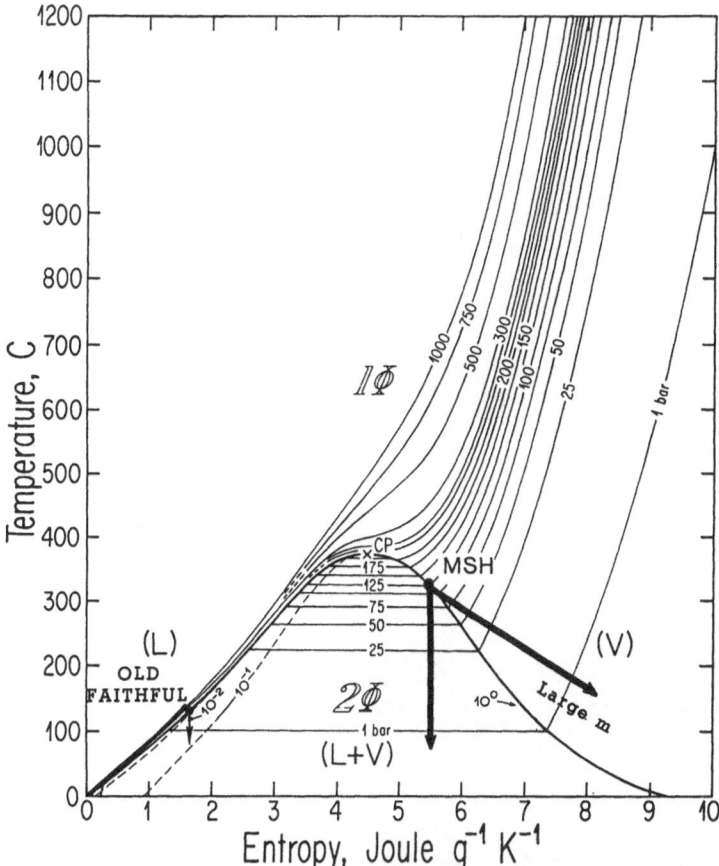

FIG. 29. Temperature-entropy diagram for H_2O with isobars. On the left, an isentrope for an eruption of Old Faithful is shown; on the right, two isentropes for eruptions from the assumed Mount St. Helens (MSH) initial conditions are shown. Pure steam would have condensed during isentropic expansion, as indicated by the vertical line. Heavy mass loading of the steam by hot particles increases the mixture entropy and, by transfer of heat to the steam, increases the entropy of the steam during expansion. Condensation during decompression is thereby prevented, as indicated by the arrow pointing to the lower right.

zone to emphasize that the blast followed channels in the topography. Surrounding this region and marking the limits of the devastated area is the *singed zone*, a zone in which trees were left standing but were singed by the heat of the blast as it became positively buoyant and lifted from the ground into the atmosphere[37].

This devastation was caused by the eruption of more than 10^{14} g of magma, hot water, and entrained glacier ice and trees. The vent through which the material emerged covered a large fraction of the north side of the mountain. Available evidence allows

192

FIG. 30. Photographs showing (a) damage to heavy logging equipment during the May 18 lateral blast, and (b) damage to trees. Photos by H. and S. Kieffer.

many theories, and geologists do not even agree on initial and boundary conditions for the flow[38]. I describe below my model for the blast, a model that emphasizes the role of gas expansion and nozzle flow[37]. Given a plausible set of simplifying assumptions, this model attempts to define the flow characteristics and to correlate these predicted characteristics with features in the devastated area: tree directions; the transition from direct to channelized blast zones; measured velocities and temperatures; and the general shape of the devastated area.

B. A simple nozzle model

In my model, the lateral blast is simplified to the problem of the eruption of a pseudogas from a reservoir under pressure into an atmosphere at lower pressure. The

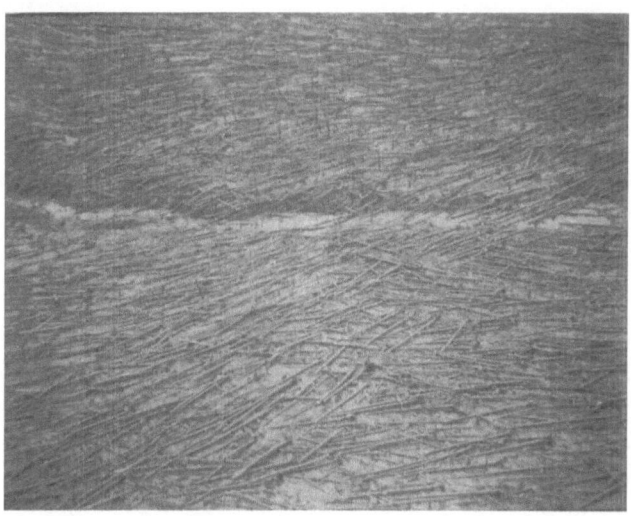

FIG. 31. Typical pattern of tree blowdown. Note that the tops of the trees and most of the small limbs are missing. The contrast in this photo is low because volcanic ash mantled trees and slope when the photo was taken shortly after May 18, 1980.

thermodynamic properties and reservoir geometry are also simplified accordingly (Fig. 28c). The fluid is defined by its initial (average) pressure, P_0; temperature, T_0; and mass ratio, m, of solid to vapor phases. The reservoir is assumed to have resembled a converging nozzle whose exit plane (vent) was the landslide scarp left on the north face of the mountain by the removal of the avalanche material.

Although the model can easily be scaled both geometrically and thermodynamically, one set of plausible values for initial conditions demonstrates the features of the model: an average reservoir pressure of 12.5 Mpa (125 bars, the pressure appropriate to 650 m of rock overlying the reservoir), an initial temperature of 600 K (327 °C), and a mass ratio of rock to steam of 25:1. The initial temperature assumed may seem surprisingly cool if one only associates volcanic eruptions with red-hot, incandescent magma. However, many so-called "volcanic" eruptions, such as the one shown in Fig. 27, are not driven by magma, but by heated water ("phreatic eruptions"), or by a mixture of heated ground water and magma ("phreatomagmatic eruptions"). The detailed nature of the volcanic gases driving the eruption is ignored here because the large-scale features of the fluid mechanics are probably not sensitive to the gas source. The chosen temperature happens to be the saturation temperature of pure water at 12.5 Mpa (125 bars), and it is a reasonable number to assume *a priori* if one believes that geothermal waters heated to saturation conditions drove the eruption (and that the badly fractured mountain edifice could not sustain any overpressure in the eruptive fluid). The temperature of 600 K can also be thought of as an average temperature for a complex mixture which, after it had traveled only a short distance, contained material ranging

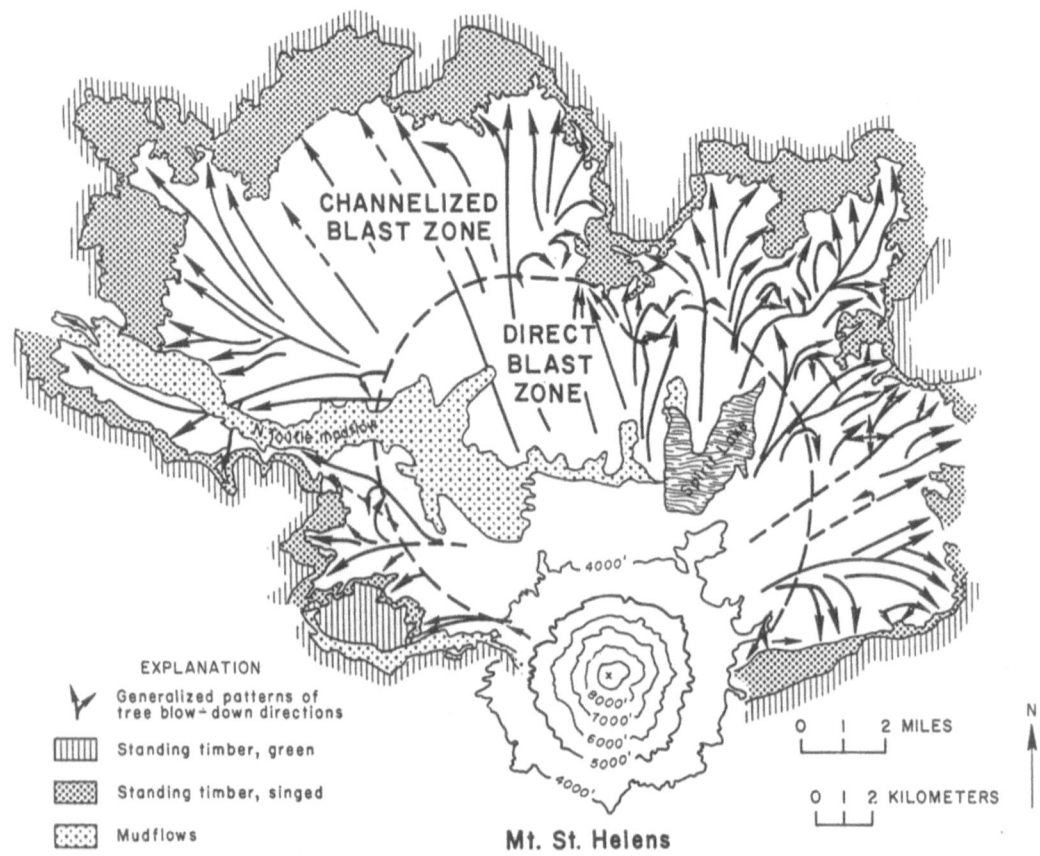

EXPLANATION

Generalized patterns of
tree blow-down directions

Standing timber, green

Standing timber, singed

Mudflows

Mt. St. Helens

FIG. 32. Map showing generalized direction of tree blowdown, with approximate extent of direct, channelized, and singed zones.

from the melting temperature of the magma (~ 950 °C) to the freezing temperature of glacial ice and snow entrained in the flow. Reasonable changes in assumed initial pressure, temperature, and solid-to-mass ratio do not qualitatively alter the conclusions. Atmospheric pressure is taken as 0.87 bar. For scale, the vent diameter is taken as 1 km, the approximate width of the scar left by the avalanches. The eruption is assumed to be centered at 2,135 m (7,000 ft), and the centerline of the flow is oriented about 5 degrees east of north to match the overall direction of the flow field. These are the only variables in the model -- there are no arbitrary fitting parameters.

Nozzles operating at pressure ratios much greater than about 2:1 are supersonic. At the ratio of 125:0.87 assumed above for Mount St. Helens, the emerging flow should have been highly supersonic (refer to Fig. 2; see Fig. 33 for a more detailed diagram and nomenclature). The most important dynamic parameter of the erupting fluid is its sound speed -- 105 m/s for the reservoir fluid postulated above, according to pseudogas calculations. This sound speed is about 1/3 of the value of the atmospheric sound

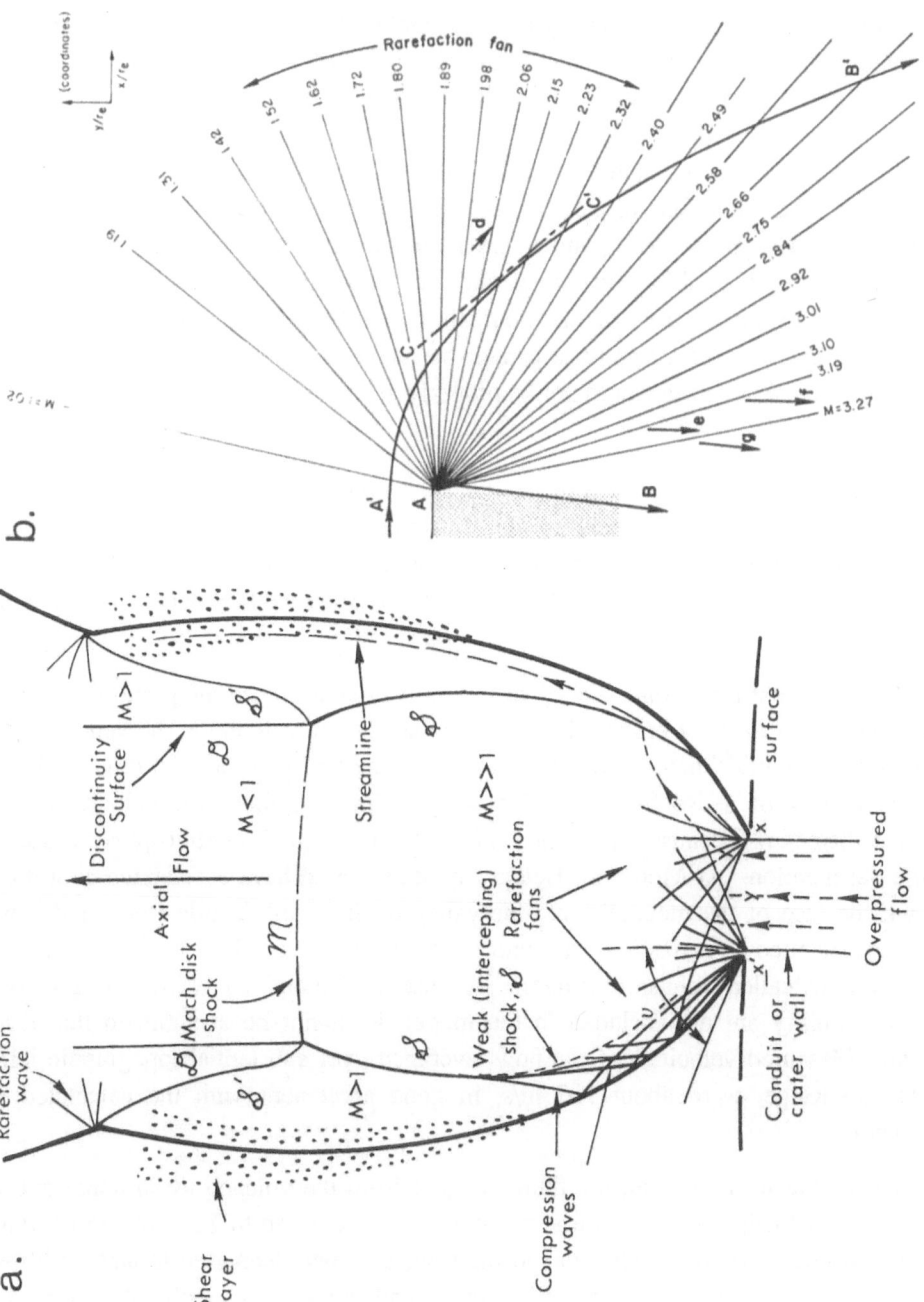

FIG. 33. (a) Schematic diagram of the structure of an underexpanded supersonic jet[39]. The flow leaves the nozzle through plane xx'. Further discussion of this figure is given in Ref. 37, p. 388. (b) Detail of a corner rarefaction for conditions appropriate to the Mount St. Helens model. This could be interpreted as a map view of the flow pattern

from the east corner of the vent (where the walls of the volcano are projected by the stippled pattern). The (mathematical) characteristics of the flow are the thin lines that radiate from the corner A of the vent. They are labeled according to the Mach number, M, of the flow as it crosses the characteristics. AB and A'B' are streamlines of the flow; CC' is a tangent to A'B'. The small arrows d, e, f, and g represent calculated local directions of flow as they would be recorded in the directions of tree blowdown. The arrows e, f, and g are particularly significant, because if these are assumed to lie along flow streamlines and are extrapolated linearly backwards, a flow source significantly in front of the mountain would be inferred. In compressible flow, linear extrapolation of streamlines cannot be made, because the curvature of the flow through expansion waves would not be properly accounted for.

---

speed. Therefore, the flow field of the volcanic pseudogas can be *internally supersonic*, but still *subsonic* with respect to the surrounding atmosphere. Thus, there is no contradiction between the postulated supersonic flow and the notable absence of atmospheric shock waves during the lateral blast (see eyewitness accounts in Ref. 35 and Ref. 37).

Consider first the initial velocity of the fluid. According to the proposed model, the fluid would accelerate from rest in the reservoir to sonic velocity at the vent -- 100 m/s. Laboratory studies[40] have shown that, in the absence of gravitational effects, the flow-front velocity of dense fluids remains at approximately the sonic velocity for many source diameters, because entrainment of the light surrounding atmosphere causes very little deceleration. At Mount St. Helens, the flow would have accelerated as it dropped down the face of the mountain into the valley of the North Toutle River, but it would have decelerated as it rose back up into the high country north of the Toutle River (the region now called Johnston Ridge). Because the combined effects of gravity and compressibility are not included in the model, it cannot be accurate to this level of detail. Measured velocities of the flow, averaged over substantial topographic relief to Johnston Ridge, were about 100 m/s, in good agreement with the calculated sonic velocity.

According to the model, the fluid emerged from the volcano as an under-expanded supersonic jet (Fig. 34). The pressure would have decreased to 7.5 Mpa (75 bars) as the fluid accelerated from the reservoir to the vent, and then decreased to ambient pressure through a series of complex rarefaction waves and shock waves within the jet outside of the volcano, as shown schematically in Fig. 33a and in detail in Fig. 34. Because of the high pressure in the jet as it left the vent, it would have spread laterally through a characteristic angle known as the *Prandtl-Meyer angle* (Fig. 33b). For the initial conditions postulated, the Prandtl-Meyer angle is 96 degrees. Thus, flow that initially was directed northward by the geometry of the vent would have diverged to the east and west; note this type of expansion beyond the east-west line in Fig. 32. The predicted

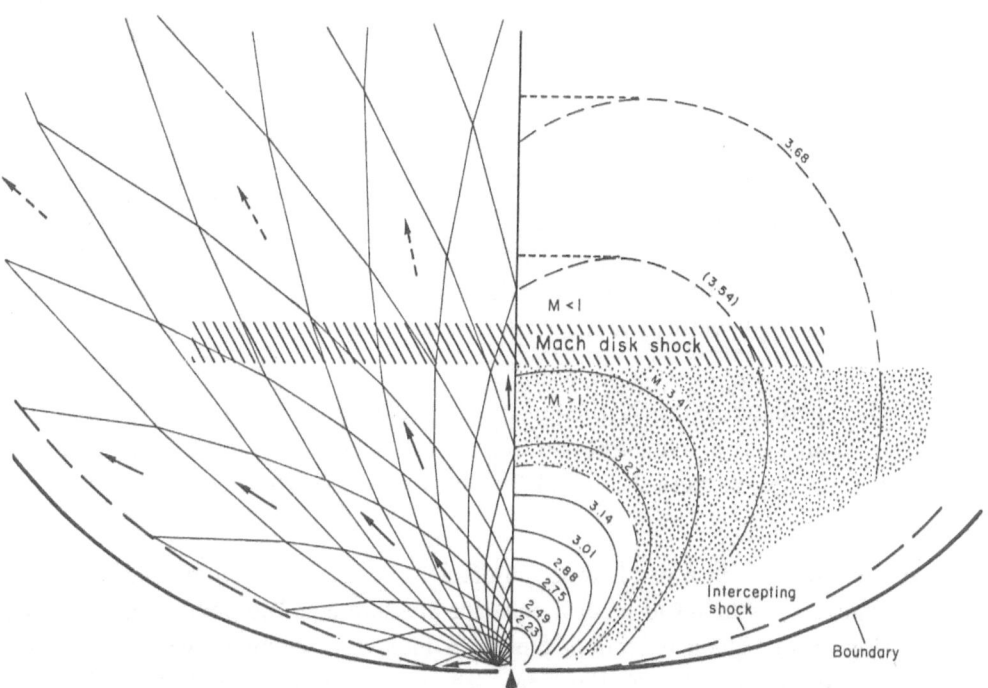

FIG. 34. Map of the flow field according to the present model of blast dynamics[37]. To ease the numerical computations, the exit Mach number of the flow is assumed to have been 1.02, instead of the sonic Mach number 1.00. All length dimensions are normalized by the vent diameter. The model is symmetric about the axis of the vent and is split into two halves here for conciseness. On the left, the characteristics (computed by hand) are shown as thin lines radiating from the corner of the vent. The boundary of the flow is assumed to have been at constant pressure (0.87 bar). The peripheral intercepting shock formed by reflection of the expansion waves from this boundary is shown as a dashed line. Note how reflection of expansion waves deflects the boundary of the flow away from its original expansion angle of 96 degrees. Flow directions are shown by representative arrows, solid within the zone where the model is strictly valid, dashed in the zone where the model is extrapolated across shock waves. On the right, contours of constant Mach number M and, therefore, constant pressure (P/P_0), temperature (T/T_0), and density (ρ/ρ_0) are shown. Velocities are given implicitly by the Mach numbers. Each contour is labeled by the value of the Mach number, M. From the innermost contour outward, values of M, P/P_0, T/T_0, and ρ/ρ_0 are given. In the supersonic region, these values are, respectively, (2.23, 0.087, 0.91, 0.095); (2.49, 0.047, 0.89, 0.053); (2.75, 0.025, 0.87, 0.029); (2.88, 0.018, 0.86, 0.021); (3.01, 0.013, 0.85, 0.016); (3.14, 0.009, 0.83, 0.011). The values extrapolated into the subsonic zone were used by the author in Ref. 37 to extrapolate the flow density to the singed zone, but should be ignored, because a more realistic assumption is that the flow returns to atmospheric pressure. The area covered by a stippled pattern is a core in the

flow that is at subatmospheric pressure. Downward curvature of the outer contours near the axis of the flow is probably an artifact of the grid size used in the numerical solution, and the likely contour shape is shown by short dashed lines. A computer model that produced a somewhat smaller supersonic zone, because the flow boundary was assumed to be inviscid rather than viscous (as in the above model), was run by R.A. O'Leary, Rocketdyne, for the author. Details can be found in Ref. 41. Differences between the two models are not significant in terms of our lack of knowledge of the real complexities of the eruption; e.g., material emerging from two moving landslides instead of from a single vertical vent.

flow zone is superimposed on the map of the devastated area in Fig. 35. I suggest that the devastated area has a southern boundary that actually curves south of an east-west line near the volcano because the initial Prandtl-Meyer expansion drove gas around in these directions.

In an under-expanded supersonic jet, rarefactions crisscross the flow and reflect off the flow boundary, assumed to be at a constant pressure equal to ambient atmospheric pressure (Fig. 33a). Upon reflection, they turn into weak compressive shocks called "intercepting" or "barrel" shocks (Fig. 33a). The reflection of the rarefactions from the flow boundary turns the diverging flow back toward a more axial direction. I suggest that these reflections are responsible for focusing the direct blast zone so strongly to the north (Figs. 32 and 35) and for limiting the extent of east-west devastation.

The fluid inside of the jet expands and accelerates as it passes through the expansion waves -- obtaining, according to the model, a Mach number of more than 3 on the centerline and velocities in excess of 300 m/s. Internal velocities can be locally higher than the flow-front velocity because of the internal rarefaction and shock waves. Pressure, temperature, and density decrease through the expansions. The pressure behavior is particularly interesting and illustrates the nonlinearity of the supersonic expansion process: as the fluid expands, the pressure decreases *below* atmospheric pressure, and a large zone of subatmospheric pressure develops inside the supersonic zone (see the shaded area in Figs. 34 and 35). The existence of such a low-pressure core has some interesting volcanologic implications; for example, plastic components on vehicles in or near this part of the devastated area were degraded by the formation of large vapor bubbles[42]. Laboratory studies demonstrated that the vapor formation was caused by exposure of the plastic to high temperatures during the blast. Efforts to duplicate the degradation by heating similar plastics in the laboratory under atmospheric pressure produced general similarities, but failed to reproduce the large size of the bubbles found on the components from the vehicles. I speculate that bubbles may have grown unexpectedly large because the external pressure was temporarily lower than atmospheric in the supersonic core of the lateral blast.

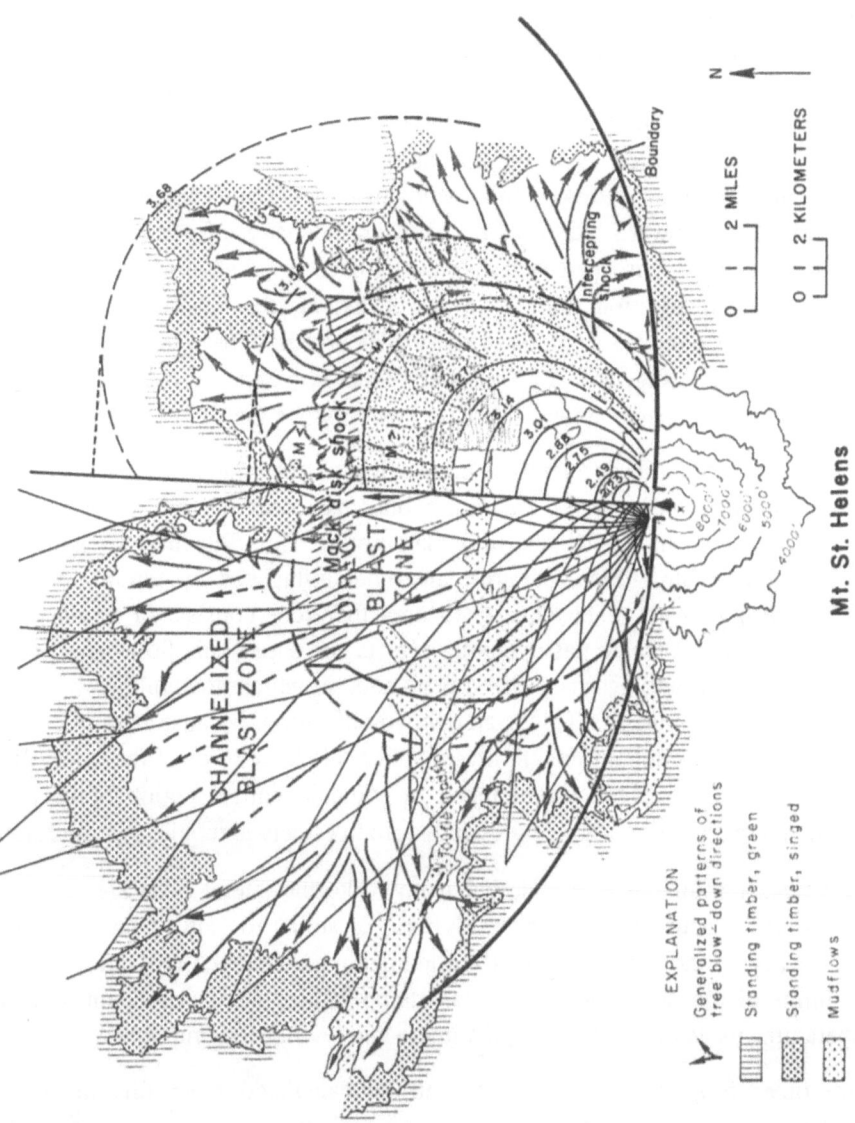

FIG. 35. The model of Fig. 34 superimposed on the map of the devastated area. The coincidence of the Mach disk with the direct blast zone boundary in the northerly direction suggested to the author that the direct blast zone was a zone of strong supersonic flow.

In an under-expanded jet, the intercepting shocks strengthen within the flow and coalesce across it into a strong shock standing perpendicular to the axis of the flow. This shock is called the *Mach disk* (Fig. 33a). As gas flows through the shock, it decelerates from supersonic to subsonic conditions; to a first approximation, the pressure on the downstream side of the Mach disk is atmospheric. Inertia of the heavy debris entrained by the blast at the Mach disk (the overturned logging vehicle in Fig. 30a and the debris around it give some impression about the size of the debris load near the Mach disk), would propel the particulate matter through a "gas shock", so that the Mach disk should, in this geologic case, be thought of as a Mach-disk zone, perhaps of the order of 1 km in thickness. As the fluid decelerates into the subsonic zone downstream of the Mach disk, flow velocities decrease, pressure rises from subatmospheric back toward atmospheric, and the density of the fluid increases. According to the calculations, the Mach disk would have stood about 11 km north of the vent. The calculated position and, to a lesser extent, the position of the lateral intercepting shocks, coincide roughly with the boundaries between the direct and channelized blast zones (Fig. 35). I propose that these two zones correspond roughly to the boundary between supersonic and subsonic flow regimes within the lateral blast.

Because of the dramatic deceleration of the flow at the Mach disk, gravity, which was not a dominant force within the direct blast zone, dominated flow mechanics outside of this zone -- in the channelized blast zone. Thus the flow streamlines, as indicated by the tree blow-down patterns, are more influenced by topography in the channelized, subsonic zone. The devastated area therefore consists of two parts: the inner direct blast zone in which gas dynamics effects and supersonic flow were probably dominant, and the surrounding channelized blast zone in which downhill flow driven by gravity was probably dominant. This is an oversimplification, because both effects were probably important throughout much of the devastated area (e.g., the most highly supersonic zone and the Mach disk happen to coincide with a region of very steep topography), and quantitative modeling including both effects is required in the future.

Temperatures throughout a particle-laden flow like the lateral blast are remarkably high and uniform because of the high mass ratio of solids to vapor (this is the effect of $\gamma \sim 1$; see Fig. 4 and Eq. 2). Calculated temperatures changed only from 600 K to 480 K at limits of the devastated area; these temperatures are in excellent agreement with temperatures measured in the deposits immediately after the eruption[35].

Several other properties of the blast can be calculated from this model. For example, the maximum mass flux is calculated to have been 10^4 g/s/cm^2, and the thermal flux to have been 2.5 MW/cm^2. The total energy of the blast was 24 Mt, of which 7 Mt was dissipated during the blast itself, and the remaining 17 Mt was dissipated during the almost simultaneous condensation of steam in the blast and the subsequent cooling of steam and rock to ambient temperature in the weeks following May 18.

As mentioned above, the supersonic flow model for the lateral blast has been controversial. Nevertheless, features analogous to those eroded into the surfaces of supersonic reentry vehicles have been found in the erosion surface under the blast deposits[43].

C. Comparison of Mount St. Helens with the Saturn V rocket

The magnitude of the blast can be impressed upon one's imagination, and the true scale of nozzles in geology can be appreciated, by comparing the Mount St. Helens blast with a Saturn V F-1 liquid-oxygen/kerosene motor (Fig. 36). The mass flux per unit area at the exit of an F-1 is about 25 g/s/cm^2; that of the lateral blast was 240 times as great. The power per unit area of the F-1 motor is approximately 0.8 MW/cm^2; that of the lateral blast was three times greater. The Saturn V power is delivered over five rockets covering roughly 50 m^2; the power at Mount St. Helens flowed out of a vent more than 2,000 times this area. The total power of the five Saturn V motors is about 4

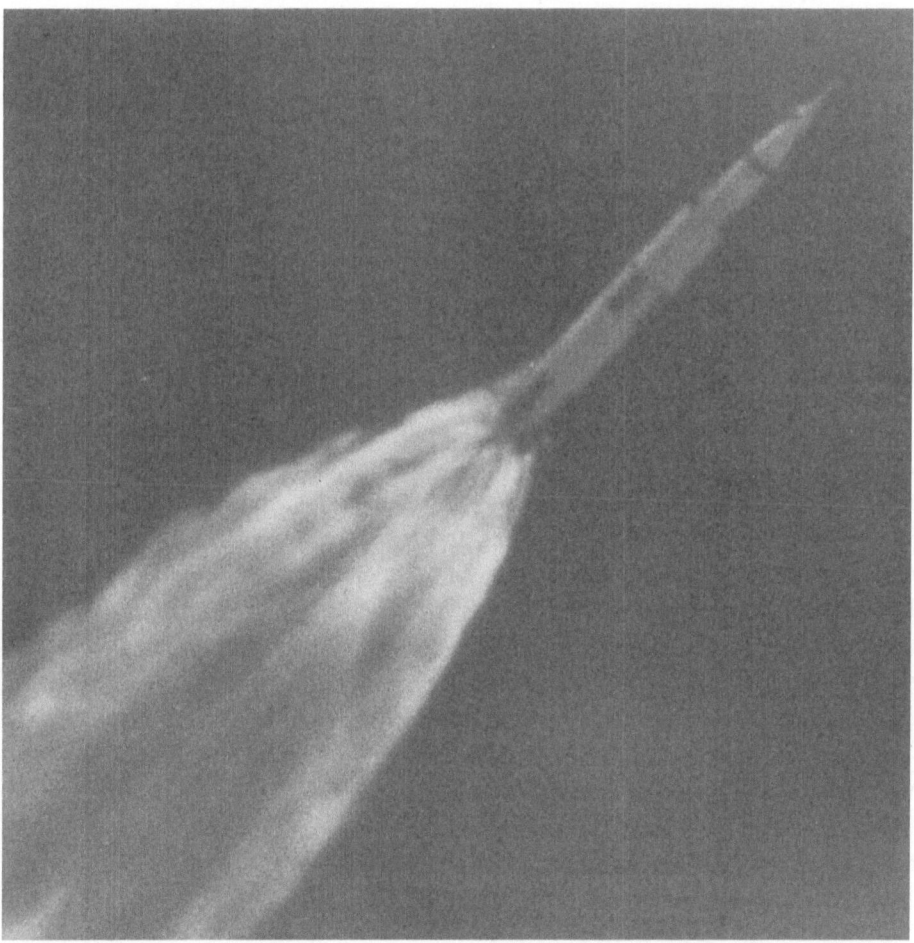

FIG. 36. A comparison of the power of Mount St. Helens with that of a Saturn V F-1 rocket engine is given in the text. Five F-1's provided the power needed to launch the Apollo spacecraft. The photograph shows Apollo 10 shortly after launch from the Kennedy Space Center on 18 May, 1969 (courtesy of NASA).

$\times 10^5$ MW; that of the blast was nearly 16,000 times greater. The thrust of the Saturn V is 7.5 million pounds (3.3×10^7 N); that of the blast was nearly 10^5 greater. The lateral blast of May 18, 1980, was indeed an awesome event by both geologic and fluid dynamical standards.

VI. Perspectives

In the discussion of Crystal Rapids and Old Faithful Geyser, I have pointed out specific directions for future research. At Mount St. Helens, unequivocal re-creation of the fluid dynamics of the lateral blast may be difficult in spite of the fact that it is the best-documented violent volcanic eruption in recorded history. The observational problems inherent in geologic research, and particularly in the monitoring of geologically rare events, are enormous. Nevertheless, the evolution of fluid dynamics in geology over the past few decades has been rapid; viz., the pioneering work of Wilson and Sparks and their colleagues and students[44].

It is appropriate to conclude with the thought that the development of modern space-craft, in which Hans Liepmann and his colleagues have been so involved, has led to one of the most exciting discoveries of modern times -- the existence of erupting volcanoes on another planet[45]. Even as we struggle to document and understand the geological physics of fluids in our world, we have already discovered new puzzles in fluid dynamics on other worlds.

Acknowledgements

The studies described here have extended over 12 years and have involved many colleagues to whom I owe thanks, including those I have referenced or mentioned as "private communication" in this article, and numerous others who could not be referenced because of space limitations: colleagues who work within the National Science Foundation, the U.S. Geological Survey, and the Bureau of Reclamation to help fund the work; and Eugene Shoemaker and Bradford Sturtevant, who interested me in, and taught me, fluid dynamics as a student from the text by Liepmann and Roshko. Finally, special thanks are owed to a supportive husband who has encouraged all my work, and to our son -- who proved a valuable field assistant as early as 1976 at the age of 10 (on many Yellowstone field trips by providing hot dogs and comics to mom, who was waiting yet another few hours for a geyser to erupt), and in 1987 could still be found enjoying helping mom carry equipment in and out of the Grand Canyon.

References and footnotes

1. Approved for publication by Director, U.S. Geological Survey, January 23, 1986.

2. For example, see H.W. Liepmann and A. Roshko, *Elements of Gasdynamics* (Wiley, New York, 1957). For the sake of brevity, references in this paper are

limited. The author has published detailed analyses of the three geologic problems discussed here in the references cited below, and the reader will find references to other relevant work in those papers.

3. This analogy was apparently first described by D. Riabouchinsky, C. R. Acad. Sci. **195**, 998-999 (1932). For a more accessible reference, see *Modern Developments in Gas Dynamics*, edited by W.H.T. Loh (Plenum, New York, 1969), pp. 1-60.

4. For example, see P.A. Thompson, *Compressible-fluid Dynamics* (McGraw-Hill, New York, 1972), pp. 517-531.

5. In the spirit of emphasizing the similarity of the various flow fields discussed in this paper, the word "nozzle" will be used interchangeably with the words "flume", "channel", and "conduit", and the word "contouring" will be used interchangeably with the word "eroding".

6. "Low" pressure ratio in this context means that the reservoir pressure is less than about 2 times atmospheric pressure; see, for example, the tables of isentropic flow variables given in M.J. Zucrow and J.D. Hoffman, *Gas Dynamics* (Wiley, New York, 1976).

7. Although schematic illustrations of the structure of supersonic gas jets can be found in most textbooks on gas dynamics, collections of actual photographs are rare. One such collection is E.S. Love and C.E. Grigsby, NACA RM L54L31 (1955).

8. A "small" head difference means that the elevation difference between the two reservoirs should not exceed approximately one-third of the head of the source reservoir.

9. Collections of illustrations of subcritical flow accelerating to supercritical flow in a converging-diverging channel are rare. Some examples can be found in E. Preiswerk, NACA TM 935 (1940). Illustrations of supercritical flow in converging and diverging channels can be found in the four papers in High-velocity Flow in Open Channels: A Symposium, Paper 2434, Trans. ASCE **116**, 265-400 (1951).

10. Reviewed in S.W. Kieffer, J. Geophys. Res. **82**, 2895-2904 (1977).

11. S.W. Kieffer and J. Delany, J. Geophys. Res. **84**, 1611-1620 (1979).

12. K. Richards, *Rivers: Form and Process in Alluvial Channels* (Methuen, London, 1983), p. 58.

13. As an example of the relative magnitudes of Froude and Mach numbers, consider order of magnitude estimates for Old Faithful and for the Mount St. Helens lateral blast. At Old Faithful, the exit velocity is ~ 80 m/s (see text, Section IV D). As

the hottest fluid (116–118 °C) ascends through the conduit to the exit plane, it becomes a two-phase mixture with about 4 weight percent vapor, for which the equilibrium sound speed is about 57 m/s at 0.8 bar atmospheric pressure at the elevation of Old Faithful. This gives a Mach number of ~1.5, indicating that compressibility effects are important. An internal (densimetric) Froude number for the jet of Old Faithful can be calculated (see Section IV D). For Old Faithful, I take nominal values of jet velocity $= u = 80$ m/s, jet density $= \rho_o = 11.2$ kg/m^3 (decompression of 116 °C water isentropically to 0.8 bar, 93 °C, 4 percent vapor), atmospheric density $= \rho_a = 0.7$ kg/m^3, and an equivalent axially symmetric conduit diameter $= D = 1.1$ m. With these parameters, the square root of the densimetric Froude number (which is the value to be compared with a Mach number) is 25. The jet is negatively buoyant because $\rho_o > \rho_a$. An internal Froude number for the Mount St. Helens lateral blast, considered as an incompressible density flow on an inclined plane, can be calculated from $Fr = u/(g'd \cos \theta)^{1/2}$, where $g' = g(\rho_a - \rho_o)/\rho_o$ (the absolute value of this quantity is taken); ρ_o is the density of the jet; ρ_a is the density of the atmosphere, assumed uniform; d is the flow thickness; and θ is the slope angle. For nominal parameters, I take a flow density of 100 kg/m^3 ($g' = 9.7$ m/s^2), $\theta = 11$ degrees, $u = 100$ m/s, $d = 100$ m. For these parameters, $Fr \sim 3.2$. This represents a minimum estimate, because internal flow velocities may have been greater by a factor of two to three. Note that the observed velocity of the front of the blast (100 m/s) is nearly identical to the sound speed of a pseudogas laden with solid fragments at a mass ratio of 25/1 (see Ref. 37), so that M ~ 1. These order-of-magnitude estimates demonstrate that quantitative models for the flow fields of Old Faithful and the Mount St. Helens lateral blast must eventually consider both compressibility and gravity effects.

14. The values of constriction used to obtain Fig. 6 were inferred by measuring the width of the surface water in the photo series of the 1973 U.S. Geological Survey Water Resources Division, one of which is shown in Fig. 5. From measurements of surface width, the constriction at Crystal Rapids in Fig. 5 is 0.33, and it plots as the left-most block in Fig. 6. However, for purposes of hydraulic modeling later in the discussion, it is necessary to assume an idealized cross section for the channel. A rectangular cross section is assumed. In this simplification, the "average" constriction used for modeling is generally less than that measured from air photos, because the shallow, slow flow across the debris fan, which shows in air photos of the water surface but accounts for only a small fraction of the total discharge, is ignored. In the case of Crystal Rapids, the model value is 0.25, so the reader should be alerted to this change in "shape parameter" when the modeling calculations are discussed. By either criterion used to determine the shape parameter, the channel at Crystal Rapids was more tightly constricted than at the older debris fans that formed before Glen Canyon Dam was emplaced.

15. Because of the common usage of cfs in hydraulics and by river observers, volume flow rates are given in both metric and English units throughout this paper.

16. Details given in S.W. Kieffer, J. Geol. **93**, 385-406 (1985).

17. The backwater above Crystal Rapids extends as much as 3 km upstream and is affectionately dubbed "Lake Crystal" by river runners.

18. The past tense is used in this discussion because the events of 1983 modified Crystal Rapids, and these calculations are not appropriate to its current configuration.

19. The calculations described here attribute all changes in flow regime to lateral constriction, because of the assumption of constant specific energy. In all rapids there are changes in bed elevation that affect the total and specific energy of the flow and, therefore, affect the transition from subcritical to supercritical conditions. The words "subcritical" and "supercritical" as used in this section therefore apply to a large-scale condition of the rapid, not to local details, because these additional effects are not accounted for. Most rapids are weakly supercritical because of changes in bed elevation, even in the regime called "subcritical" in this section. There are also substantial small-scale irregularities in bed topography, such as ledges and rocks, that cause local supercritical flow. The behavior of the river in flowing around such obstacles is not included in this generalized discussion.

20. S.W. Kieffer, J. Volc. Geoth. Res. **22**, 59-95 (1984).

21. C.O. McKee, D.A. Wallace, R.A. Almond, and B. Talais, Geol. Survey, Papua New Guinea Mem. 10, 63-84 (1981).

22. F. Birch and G. Kennedy, in *Flow and Fracture of Rocks*, Geophys. Monogr. Ser. **16**, 329-336 (1972).

23. Summarized in S.W. Kieffer and J. Westphal, EOS: Trans. Amer. Geophys. Union **66** (46), 1152 (1985); details in preparation.

24. D.E. White, Amer. J. Sci. **265**, 641-684 (1967).

25. S. Fujikawa and T. Akamatsu, Bull. Japan Soc. Min. Eng. **21**, 223-230 and 279-281 (1978).

26. For example, see L. Biasi, A. Prosperetti, and A. Tozzi, Chem. Eng. Sci. **72**, 815-822 (1972).

27. W. Hentschel, master's thesis (Göttingen, 1979); also Fortsch. Akustik, DAGA'80, Berlin, p. 415-418.

28. The geometry of the conduit determines whether the conduit should be considered open-ended or closed-ended. Since no probe has demonstrated a large reservoir at the bottom of the conduit, it is treated here as a closed pipe. Resonant frequencies of an open pipe would be half of those calculated from this formula.

29. J.S. Turner, J. Fluid Mech. **26**, 779-792, 1966; C.J. Chen and W. Rodi, *Vertical Turbulent Buoyant Jets -- A Review of Experimental Data* (Pergamon Press, New York, 1980).

30. G.B. Wallis, *One-Dimensional, Two-Phase Flow* (McGraw-Hill, New York, 1969), and references cited therein.

31. F.J. Moody, Trans. ASME J. Heat Transfer **87**, 134-142, 1965, as reviewed by Wallis (Ref. 30, p. 48).

32. H.K. Fauske, ANL Report 6633, 1962, as reviewed by Wallis[30].

33. S. Fuller and J. Schmidt, *Yellowstone in Three Seasons* (Snow Country Publications, Yellowstone National Park, Wyoming, 1984).

34. R.C. Leet, J. Geophys. Res., in press; contains review of other models for volcanic tremor.

35. Many detailed articles about Mount St. Helens can be found in P.W. Lipman and D.R. Mullineaux, eds., *The 1980 Eruptions of Mount St. Helens, Washington*, U.S. Geological Survey Prof. Paper 1250 (U.S. Gov't. P.O., Washington, 1981).

36. J.G. Moore and W.C. Albee, in Ref. 35.

37. S.W. Kieffer, pp. 379-400 in Ref. 35.

38. See, for example, M. Malin and M. Sheridan, Science **217**, 637-640, 1982; J. Eichelberger and D. Hayes, J. Geophys. Res. **87**, 7727-7738; or J. Moore and C. Rice, in *Explosive Volcanism: Inception, Evolution, and Hazards*, edited by F.M. Boyd (Nat. Acad. Press, Washington, 1984), Chap. 10.

39. JANNAF [Joint Army, Navy, NASA, Air Force] Handbook of rocket exhaust plume technology, Chemical Propulsion Information Agency Publication 263, Chap. 2 (1975).

40. S.W. Kieffer and B. Sturtevant, J. Geophys. Res. **89**, 8253-8268 (1984).

41. S.W. Kieffer, in *Explosive Volcanism: Inception, Evolution, and Hazards*, edited by F.M. Boyd (Nat. Acad. Press, Washington, 1984), Chap. 11.

42. M.J. Davis and E.J. Graeber, EOS: Trans. Amer. Geophys. Union **61**, 1136 (1980).

43. S.W. Kieffer and B. Sturtevant, Erosional furrows formed during the lateral blast at Mount St. Helens, May 18, 1980, submitted to J. Geophys. Res.

44. Comprehensive reviews of this work can be found in L. Wilson and J.W. Head III, Nature **302**, 663-669 (1983), and in R.S.J. Sparks, Bull. Volcanol. **46** (4), 323 (1983).

45. Review articles on Ionian volcanism can be found in D. Morrison, *Satellites of Jupiter* (U. Arizona Press, Tucson, 1982), including S.W. Kieffer, Chap. 18; see also Ref. 41.

Lecture Notes in Mathematics

Lecture Notes in Physics